Pandora's Baby

BOOKS BY ROBIN MARANTZ HENIG

The Myth of Senility

Your Premature Baby

How a Woman Ages

Being Adopted (with David Brodzinsky and Marshall Schechter)

A Dancing Matrix

The People's Health

The Monk in the Garden

Pandora's Baby

Pandora's Baby

HOW THE FIRST TEST TUBE BABIES SPARKED THE REPRODUCTIVE REVOLUTION

ROBIN MARANTZ HENIG

HOUGHTON MIFFLIN COMPANY
BOSTON NEW YORK 2004

Visit our Web site: www.houghtonmifflinbooks.com.

Library of Congress Cataloging-in-Publication Data

Henig, Robin Marantz.
Pandora's baby : how the first test tube babies sparked the reproductive revolution / Robin Marantz Henig.
p. cm.
Includes bibliographical references and index.
ISBN 0-618-22415-7
1. Fertilization in vitro, Human. 2. Human embryo — Transplantation.
3. Human reproductive technology. I. Title.

RG135.H46 2004
618.1'78059 — dc22 2003061372

Printed in the United States of America

MP 10 9 8 7 6 5 4 3 2 1

TO JEFF,
a bushel and a peck

Contents

Prologue:
Monster in a Test Tube 1

Part One: *Ex Ovo Omnia*

1. Room Temperature 19
2. The Dance of Love 26
3. Laughingstock 56
4. Out of Control 64
5. Fits and Starts 78
6. Laboratory Ghouls 87

Part Two: The Modern Prometheus

7. Toward Happily Ever After 95
8. Baby Dreams 104
9. Science on Hold 118
10. The First One 133
11. A Baby Clone 142
12. Hang On 150

Part Three: Test Tube Death Trial

13. Fooling Mother Nature 155
14. Pandora's Baby 170
15. Normality 173
16. Prometheus Unbound 177
17. Verdict 193

Part Four: Not Meant to Be Known

18. Right to Life 201
19. Opening Pandora's Box 217
20. Tables Turned 229
21. From Monstrous to Mundane 233
22. Pandora's Clone 245
23. Mixed Blessings 261

Notes 269
Selected Readings 303
Acknowledgments 308
Index 310

Ex ovo omnia.

From an egg, everything.

—William Harvey,
De Generatione Animalium (1651)

Prologue

Monster in a Test Tube

> Frightful must it be, for supremely frightful would be the effect of any human endeavor to mock the stupendous mechanism of the Creator of the world.
>
> — Mary Shelley, *Frankenstein, or The Modern Prometheus* (1818)

On a cool fall morning in 1973, Doris and John Del-Zio arrived with her luggage at New York Hospital. It was a familiar routine for the Florida couple; Doris had been a patient there before. On three earlier occasions her Manhattan infertility specialist, William Sweeney, had tried surgically to remove obstructions in Doris's blocked fallopian tubes. The first surgery worked and Doris became pregnant, but three months along she had a miscarriage. The second and third surgeries had no effect at all. Neither did attempts at artificial insemination using her husband's sperm, not after the first insemination, nor the second, nor the third. Month after month after disappointing month, Doris Del-Zio, then approaching thirty, got her menstrual period, and each one was a stinging rebuke. Every period taunted her — *no baby, no baby, no baby* — forcing her to acknowledge that she still wasn't pregnant and probably never would be.

Maybe she and John should have just left well enough alone. Maybe God, or fate, or whatever one calls the keeper of one's destiny, meant for Doris to be content with her ten-year-old daughter,

Tammy, the child of her first marriage, and with her two college-age stepdaughters, Denise and Debbie, who lived with John's ex-wife. Maybe it was enough to have a beautiful home in Fort Lauderdale and an adoring husband, a professional man — a dentist — who had adopted Tammy and loved her as though she were his own. But Doris wanted to have John's baby, and she was ready to do almost anything to make that happen.

"Isn't there something else you can do?" Doris asked Sweeney after her third failed surgery. She was a pretty woman with brunette hair swept into a lacquered flip, and her dark eyes were sad. "They can put a man on the moon; isn't there some way scientists can figure out how to help me have a child?"

Well, yes, Sweeney conceded, a little reluctantly because of Doris's long history of infertility surgery; there was one more thing they could try. It had never been done in humans before, only in lab mice and rabbits. But if Doris was willing, he could try a new method, in vitro fertilization, or IVF, which would bypass her clogged tubes altogether. The few journalists who had written about the procedure were calling it the creation of test tube babies.

If the Del-Zios consented — and Doris took barely ten minutes to decide that this was her last, best hope — Sweeney said he would surgically remove a few of Doris's eggs and, with a collaborator who had done such things before, fertilize them with John's sperm in a glass test tube (*in vitro* is Latin for "in glass"). If one of the sperm fertilized one of the eggs, the resulting zygote — the scientific term for a fertilized egg — would be placed in an incubator at body temperature for three or four days and allowed to grow. The single cell would become two, the two would become four, the four eight, the eight sixteen, and the sixteen thirty-two. It would take about three days for the zygote to grow into the thirty-two-cell ball known as a morula and another day or so to grow into a blastocyst, a fluid-filled sphere made up of a few hundred cells. Even though it would still be smaller across than the width of an eyelash, the blastocyst would now be ready to implant itself in the uterine wall. According to the plan, then, Sweeney would have Doris return to the operating room four days after her eggs had been harvested, to introduce the minuscule blastocyst into her uterus at about the same time nature would have done so had it been given the chance. Her body, he hoped, could take it from there.

The logistics of the undertaking were tricky, made even trickier by geography. Doris's eggs would be removed at New York Hospital, an affiliate of Cornell Medical School on East 68th Street on Manhattan's Upper East Side. But Sweeney believed that the only man in New York with the experience, the interest, and the nerve to try to fertilize those eggs in vitro was a physician named Landrum B. Shettles, who worked at Columbia Presbyterian Medical Center on West 168th Street and Broadway, one hundred blocks north and on the opposite side of the city, in Washington Heights. That was where the fertilization would have to take place.

By the time Doris checked into New York Hospital to attempt IVF, she had been taking fertility drugs for more than six months to pump up the activity of her ovaries. Sweeney met with her and John to discuss the procedure and to have them read and sign all the consent forms. In addition to the usual forms, Sweeney handed the couple a one-page document that seemed almost improvised, so casual was its tone, so lacking in standard legalese: "Of our own free will and volition and with full knowledge of in vitro fertilization, . . . [we] hereby authorize Dr. William J. Sweeney to perform a laparotomy with embryo transplant and any other operation upon Mrs. Doris Del-Zio, and to employ any assistance as he may desire to assist him. We understand that there is no guarantee or assurance that a pregnancy could result. We understand that there is the possibility of complications of pregnancy and of childbirth and delivery, or the birth of an abnormal infant or infants, or undesirable tendencies or other adverse consequences."

The IVF consent form, the first of its kind used in New York — indeed, possibly the first of its kind used anywhere in the world — was dated September 11, 1973. It was typed on a manual typewriter with a smudgy black ribbon, signed by Doris in back-sloping handwriting with a black fine-point pen and by John in bigger, bolder script in blue.

On September 12, a Wednesday, Sweeney came in to the eighth-floor operating room where he had encountered Doris so many times before, made a new incision in her abdomen, and used a syringe to draw out from her ovaries what is known as follicular fluid. In a woman who has been taking fertility drugs, these ovarian secretions usually contain at least a few eggs.

Sweeney collected about one cubic centimeter of fluid, which he divided between two test tubes, adding some tissue scraped from the fallopian tubes for nourishment. Then he phoned Landrum Shettles at Presbyterian Hospital to tell him the eggs were on their way.

It was John Del-Zio's job to get them there. Nestling his wife's eggs inside his jacket pocket, safe in two corked test tubes swaddled in bubble wrap, he took the elevator down to the York Avenue exit. Then Del-Zio, a good-natured man with a reedy voice, thinning black hair, and the looks of Phil Silvers on the *Sergeant Bilko* television show, began the journey that he and his wife hoped would lead to the world's first test tube baby.

The Del-Zios may have thought they were just trying to make a baby, but in truth they were also making history. And, finding themselves in the swirl of an epoch-defining vortex, they were about to come face to face with their own true selves: part courage, part vanity; part selflessness, part greed.

But John Del-Zio did not know that yet. All he knew was that he had to get over to the West Side and up to Washington Heights. So he stood in the morning chill, peered at the traffic going north on York Avenue, and hailed a cab.

The enterprise the Del-Zios were embarked upon, in vitro fertilization, carried a slightly sinister overtone in September 1973, somewhat like the back-alley connotations of abortion before the Supreme Court's *Roe v. Wade* decision, just eight months earlier. Within a few short years, spurred mostly by scientists' ability to bring into the world test tube babies who were perfectly beautiful, normal-looking in every way, society's view would begin to change. The change was subtle, a cultural and intellectual shift so gradual as to be almost imperceptible to those living through it. But it added up to something radical indeed. A graphic demonstration of this evolution can be seen in the story of the Del-Zios, their physicians, William Sweeney and Landrum Shettles, and Raymond Vande Wiele, the chairman of Shettles's department, whose attitude toward IVF underwent a dramatic reversal over the course of a single decade. This book is about that transformation, the people who struggled through it, and the regulatory mechanisms that society put in place during that tumultuous time — mechanisms that are still in place today to guide us through our next adventures in reproductive technology.

It's an American story, even though early IVF is often associated with England, where the world's first test tube baby — Louise Brown, billed as "the Baby of the Century," her doctors hailed as heroes — was born in 1978. But the American IVF attempt involving the Del-Zios, Shettles, and Vande Wiele, which began five years earlier, reveals, perhaps better than the more familiar British story, what can happen when society faces a new and frightening technology: how it is greeted first with resistance and expectations of the worst, then with grudging permission, then with acceptance, and finally with incorporation so seamlessly into the culture that no one even notices it anymore. Because the two stories, the American and the British, took place concurrently, both are told here. And alongside the stories of the research itself are some larger, more perennial issues: the struggle between the drive to know and the drive to not know; the growth of the field of bioethics; the mechanisms by which new technologies are introduced and regulated; and the factors that motivate scientists, including altruism, personal bravura, economics, and lust for power.

It seems hard to believe today, when the procedure is so routine that it is usually covered by medical insurance, that IVF in 1973 was thought by some to threaten the very fabric of civilization. Marriage, fidelity, the essence of family; our sense of who we are and where we're headed; what it means to be human, connected, normal, acceptable; ideas about love, sex, and nurturance; the willingness to yield to the inscrutable, marvelous mystery of it all. If in vitro fertilization was allowed, some said, all the stabilizing threads would unravel.

The threads were unraveling already, of course, which is probably why IVF seemed so threatening, yet another tug at the ever-loosening weave. Feminism was a major source of the fraying. In the early seventies women were rewriting their social roles, moving out of housewifery, delaying childbearing or choosing to be childless altogether, demanding access to traditionally male domains. With the feminist movement turning motherhood into an option instead of an obligation, any proposed change in the relationship between a woman and her reproductive capability was particularly fraught. The birth control pill had already separated sex from procreation; the *Roe v. Wade* decision had already separated pregnancy from birth; no-fault divorce laws had already separated marriage vows from forever. With all these coincident changes, what would the new reproductive technology do to our perception of children as the fruit of a loving, lifelong union?

"Marriage and the family must be abolished as institutions," wrote Ti-Grace Atkinson, the author of *Amazon Odyssey* and one of the most radical feminists of the day. "And 'love' as an ideology to justify them must also go."

Even more than the institutions that supported it, the act of procreation itself was reassessed and found to be politically suspect. This was the era of Zero Population Growth (ZPG), when couples who had more than two children were viewed askance, thought to be wantonly gobbling up the planet's precious gifts. To true believers, nothing less than the fate of the earth was at stake, its environment imperiled by too many people and too much technology.

Just as ZPG was making people think twice about having babies, a growing environmental movement was making them think twice about scientific advancement for its own sake. The notion that progress almost always comes at a cost underlay one of the movement's earliest achievements, the Environmental Impact Assessment, introduced in the National Environmental Protection Act of 1970. Good social policy, the Environmental Impact Assessment made clear, denied neither the progress nor the cost but sought to balance the two in a morally responsible way.

Comfortable conventions were suddenly open to reevaluation, too. The antiwar movement, which helped lead to America's humiliating withdrawal from Vietnam in 1973, provided a new matrix for social cynicism and the belief that the government sometimes makes grave mistakes. The government's fallibility was reinforced when the Watergate scandal erupted on national television. The daily congressional hearings turned the summer of 1973 into an object lesson in how long, and how destructive, can be the shadow cast by a single too-powerful man.

In that same year the vice president resigned in ignominy over charges of graft and tax evasion; a group of enraged Native Americans laid siege for three months to a tiny South Dakota town; and an Arab oil embargo quadrupled gas prices and made Americans question their dependence on foreign oil. Into this bubbling mess of social change came scientists who wanted to create life in the laboratory. No wonder it looked like such a dangerous idea.

No one but the gods should tamper with the natural order of things. That, at least, is the moral of the parables that have been handed down for millennia, designed to quell humankind's unpredictable, irrepressible, sometimes foolhardy impulses to twist nature according to its own whims. The folly of such actions has been the point of myths and folk tales dating back to the ancient Greeks, who told the story of Prometheus to show that any attempt by a mere mortal to create life — or, more blasphemous still, to conjure a thinking, feeling, independent organism — can lead only to ruin.

Prometheus was a Titan, not a god, but the gods adored him. During the time when all the earth's creatures were being made, Zeus gave Prometheus a special task: to create Man. Prometheus took great pride in Man — some might say too much pride, the kind the Greeks called hubris. He wanted to endow his creation with a special gift, something unique to Man, something more valuable than the gifts of flight, or strength, or speed, or camouflage, which had already been bestowed upon earth's other creatures. Prometheus decided to give Man a tool that the gods alone possessed, which would enable him to fashion other tools, to provide himself with clothing, shelter, and food. He decided to give Man the gift of fire.

Prometheus stole the fire from Mount Olympus in the dead of night and carried it to earth, nestled carefully in the crook of his arm. When Zeus saw the flickering light of the flames, he knew what Prometheus had done. He concocted a brutal punishment: he had Prometheus lashed to a rock at the top of the Caucasus Mountains, and directed a vulture to tear ceaselessly at his liver. The Titan's agony never ended. The vulture was forever hungry, and every night Prometheus's liver regenerated, ready to become the next morning's meal.

The mythic hero's suffering inspired much poetry, such as Lord Byron's "Prometheus" of 1816. In that year Byron made an excursion to the French Alps with Percy Bysshe Shelley and Shelley's teenage mistress. Shelley at the time was a married man, but a few months later he became a widower when his jilted wife drowned herself in the Serpentine River. He then married his young lover, Mary Godwin, who became known as Mary Shelley.

During that trip to the Alps the weather was bad, and Byron proposed that he, Shelley, and Mary pass the time writing ghost stories. Mary, much less accomplished than the two men, was at first struck nearly dumb by the idea. But the story she put together, about a scientist she called "the modern Prometheus," eventually became the novel *Frankenstein,* a book whose impact far outlasted that of the ghost stories concocted by Byron and Shelley.

Byron's ode addressed Prometheus directly, extolling his bravery in the face of injustice.

> Thy godlike crime was to be kind;
> To render with thy precepts less
> The sum of human wretchedness,
> And strengthen man with his own mind.

Prometheus's transgression was to provide such a powerful tool, one that would "strengthen man with his own mind." Fire was, like knowledge, a double-edged gift, with the capacity for both creation and destruction. So it is with our most potent scientific discoveries. The more wonderful the accomplishments that are made possible by science, the more potentially terrible they are as well.

When the gods looked at Prometheus and Man, they saw what they themselves had wrought: the fierce attachment that inevitably grows between creator and creation. What if Man, with toolmaking and other capabilities of fire at his disposal, developed a stronger allegiance to Prometheus than to the gods? Once the ability to enhance life was bestowed on mere mortals, how could anyone maintain order or decency or restraint?

The Promethean trick of creating life is the essence of in vitro fertilization. But it entwines two extremes of life: on the one hand, life at its most fragile and natural, involving acts of love and touch and sex and generation, and, on the other hand, life at its most contrived, with surgical interventions, microscopic examinations, and lab cultures, turning the creation of a baby into a matter of technology rather than nature. In the early days of IVF, before it became a focus of widespread public attention, the debate was mostly about science: whether human sex cells and embryos could be grown successfully in a petri dish; whether fertilizing them there would lead to gross abnormalities; whether more animal research should be done before starting work on humans. But the debate soon took on a more philosophical tone. Some people argued that a human embryo, even a single-celled zygote, deserved the same respect due a human being — not because the zygote was a person, really, but because it was a potential person, containing all the DNA necessary to make a new and unique individual. "A blastocyst . . . is not humanly nothing," wrote biologist Leon Kass, one of the loudest opponents of IVF in the seventies — and today a forceful critic of cloning. "It possesses a power to become what everyone will agree is a human being."

The very notion of artificial fertilization went against the teachings of many religious groups, including the Catholic Church. "Fecundation must be carried out according to nature and through reciprocal and responsible love between a man and a woman," said the Reverend Pierfranco Pastore, a spokesman for the Vatican, shortly after Louise Brown was born in 1978. Added the Reverend Anthony Bevilacqua of the Diocese of Brooklyn: "We would not like to see the point where science dehumanizes the act of marriage."

If some of these comments sound like a rehearsal for today's debates about human cloning, that is because there are some very real parallels. Cloning today evokes many of the same responses that IVF did thirty years ago. In fact, early opponents of IVF deliberately linked the two by predicting that human cloning was what lay at the bottom of the long and treacherous "slippery slope" down which we would inevitably tumble if IVF was allowed.

Throughout recent history, scientific and cultural changes have been subjected to the slippery slope argument. People talked about the slippery slope when the first human artificial insemination was publicized in 1909, conjuring images of selective breeding and a race of illegitimate souls. They talked about the slippery slope after the first heart transplant in 1967, after the first animal-to-human transplant in 1984 and, in the summer of 2001, after the first attempt to create human embryos explicitly for research. Early cases of assisted suicide stimulated talk of a slippery slope leading to wholesale killing of the aged or infirm; early attempts at amniocentesis, of a slippery slope toward the elimination of fetuses that were trivially imperfect — or simply the "wrong" sex.

There is power in these arguments, because if we hadn't allowed those first steps — the refinement of intrauterine diagnosis, the definition of brain death, the limbo created by the heart-lung machine — the more disturbing applications could not have come to pass.

The same can be said of IVF. Scientists first had to learn how to fertilize human eggs in the lab and how to transfer them back into the womb before they could even begin to think about the scenarios that now cause so much concern: not only cloning but also preimplantation genetic diagnosis, genetic engineering of sex cells, the creation of human/animal chimeras, the culturing of human embryos as sources of replacement parts. None of these interventions could be accomplished without first perfecting the techniques of artificial fertilization and embryo transfer.

But for all that people railed against specters of where IVF might lead, the protests had an unintended and paradoxical effect: they led to less control over IVF rather than more. Early on, opponents of the procedure thought that the best way to stop troublesome science was to keep the government from financing it, and they fought against using taxpayers' money for research involving fetuses or embryos — which, by extension, included IVF. A succession of bioethics commissions reviewed these bans on government financing, and one by one the commissions recommended that the bans be lifted. But politicians, afraid of alienating the vocal antiabortion lobby, which took on IVF as its cause, generally did not want to underwrite such controversial research. So they tended to ignore each report and form a new commission in the hopes that it would reach a different conclusion. This became the pattern for the role of bioethicists in the regulatory minuet: sit on a commission, hold meetings, attend public hearings, write a report that says the research is ethically acceptable, have the report ignored, watch the next president or Congress convene a new commission. Repeat.

Even after the ban on fetal research was finally lifted, and then the ban on embryo research, the government still refused to sponsor IVF research. But the lack of federal support didn't stop scientists from working on IVF — it just forced them to do so beneath the radar. They were thus beyond the reach of the main mechanism for oversight, which was (and still is) the federal research grant and the standards it imposes on recipients. No government grants for in vitro fertilization meant that no one was forced to adhere to any standards. But entrepreneurial scientists were doing IVF anyway, bolstered by private money from infertile couples desperate for babies of their own. Many of these scientists were honorable men and women with solid reputations and the loftiest of goals. But some were motivated by the factors that drive so many innovators, scientists included: ego, curiosity, ambition, even greed. They were free agents who essentially did whatever they wanted and whatever the market would bear. Their privately funded efforts turned some aspects of IVF into a cowboy science driven by supply and demand.

Cloning is in many respects today's cowboy science; cloners are the daredevils and rogues, making claims on television and at congressional hearings that are rarely backed up with genetic proof or an actual baby. Alarmed, many politicians in the United States and elsewhere have tried hard to put cloning in its place — not by refusing to fund it, as they did with IVF, but through legislation to outlaw it altogether — whether for research or for creating a baby. They want to keep human cloning from going the way of IVF, which developed at its own pace and became part of the ordinary landscape simply because it was easier to ignore a controversial new technology than to regulate it.

Cloning resembles IVF not only in the legislature but in the laboratory as well. For both IVF and cloning, the first step is to create a human zygote in culture. But though similar in terms of laboratory technique and in terms of the intention to allow infertile couples to have biological children, cloning and IVF have some crucial differences — and we misread the lessons of IVF for today's cloning debates if we fail to see those differences.

The goal of in vitro fertilization is to mimic sexual reproduction and produce a genetically unique human being, a baby with one father and one mother. Only the locus of conception changes, after which events proceed much the way they do in a normal pregnancy. Cloning, however, disregards sexual reproduction; it mimics not the process but the end result, the human being himself. What is produced is not a new person with a unique combination of mother's and father's DNA but the identical twin, a genetic replica in every way, of a person who already exists.

Perhaps the biggest difference between IVF and cloning is the focus of our anxieties about them. In the 1970s the greatest fear about in vitro fertilization was that it might fail, leading to sorrow, disappointment, and possibly the birth of grotesquely abnormal babies. Today the greatest fear about cloning is that it might succeed.

In terms of the evolution of the species, cloning could have serious unintended consequences — far more consequences than "basic" IVF has had. As early as the 1940s, the British author C. S. Lewis warned that the net result of reproductive technology might well be not advancement but, perversely, a bizarre kind of petrification, the freezing of the world at the particular moment in time when the new technology was introduced. It would be like walking into a twenty-first-century home and finding an avocado-colored refrigerator and brown shag carpeting. That might have been the latest fashion when the owners made their first decorating decisions — but now, thirty years later, it all looks shabby and out-of-date. Something analogous could happen to the human species, said Lewis. Babies designed according to one era's fashion could become, like pine-paneled rumpus rooms, something we regret when the fashions change. And an outdated genome can't be ripped out like an old carpet.

"If any one age really attains, by eugenics and scientific education, the power to make its descendants what it pleases," wrote Lewis in the 1943 essay "The Abolition of Man," "all men who live after it are patients of that power. They are weaker, not stronger; for though we may have put wonderful machines in their hands we have preordained how they are to use them."

Lewis's warning carries an important reminder: that opening some doors and not others automatically prevents us from venturing into certain rooms. But his emphasis is slightly askew. He makes it sound as though once we set off along a particular path of discovery, we continue to make decisions that cannot be undone. More often, however, the doors we close and open along the way are like swinging saloon doors; the process does not have to happen in only one direction. Our choices have ramifications, to be sure, but the ramifications are not necessarily linear — nor are they necessarily permanent. If we seem to have enshrined the wrong fashion, we would probably have time to find ways to undo our mistake. The challenge is to achieve a balance between making reasonable choices and being so frightened of the wrong choices that we make no choices at all.

The Prometheus story has a sequel. It involves the first woman, the ancient Greek equivalent of Eve, whom the gods sent to earth in direct retribution for Prometheus's misdeed. Her name was Pandora, meaning "all gifted." The gods on Mount Olympus fashioned her with all the most alluring traits they could think of. Aphrodite gave her beauty; Hermes, persuasion; Apollo, his magnificent music.

When Pandora arrived on earth, she was carrying a beautiful and mysterious box. The box (some myths describe it as a jar) was a gift from the gods, who handed it to her and told her never to open it. But among Pandora's many gifts were some distinctly human qualities — curiosity, audacity, impetuousness, cheek — at play in a fatally flawed combination. She defied the gods' injunction and opened the box. In doing so, she unleashed all the terrors that had until then been unknown to Man, living as he did in a blissful, innocent paradise, and that would forever after cause him anguish and pain. These grievous sorrows, as one account of the Pandora myth put it, rushed from the box "in a black stinking cloud like pestilent insects — sickness and suffering, hatred and jealousy and greed, and all the other cruel things that freeze the heart and bring on old age." The release of these miseries represented an end to the golden age, a coda to mankind's idyllic childhood.

When scientists started talking, in the early 1970s, about creating a kind of Pandora's baby, a lab-fertilized egg brought into being by human technology instead of by the gods, some observers thought again about the lessons the Greeks had tried to teach. It seemed to boil down to a struggle between two competing impulses: the creative drive to understand nature versus the conservative drive to impose limits and maintain the status quo. This is the way frontier science has always been done, through the raucous to-ing and fro-ing of contradictory desires. The conflict between striving to know and wanting not to know has been with us since Eve tasted the apple, since Prometheus brought fire to mankind.

Would Pandora's baby lead to something so close to what happened in the myth, people wondered, that the only responsible thing would be to make sure it was never born? Would successful in vitro fertilization demand a reassessment of qualities so central to our humanity — our sense of doom and destiny, our understanding of who we are and where we are headed, our definitions of parents, children, love, sex, generation — that its very existence would threaten our collective soul? These questions may seem overdramatic today, unless you replace "in vitro fertilization" with "human cloning." But in the years before the first test tube baby, these questions were asked by reasonable men and women who sincerely believed that IVF might unleash a scourge of woeful possibilities that, as with Pandora's opening of her dreadful box, we would be better off having never seen.

Part One

Ex Ovo Omnia

1

Room Temperature

> The types of questions that normally arise about any new and dramatic technological procedure fall into the categories of: can man, will man, and ought man.
>
> — Willard Gaylin, "The Frankenstein Myth Becomes a Reality" (1972)

John Del-Zio ferried his wife's eggs through the braying snarl of crosstown traffic, determined to hand them over to Landrum Shettles, one of the most famous baby makers in New York. When he arrived at Presbyterian Hospital at about eleven o'clock on the morning of September 12, 1973, he asked the woman at the front desk to direct him to Shettles's office. Suddenly Shettles himself appeared in the foyer, a small bald man with the skittish manner of a shore bird. At the time, Del-Zio didn't think it odd to be meeting the illustrious physician in the lobby instead of a doctor's office, which would have been more private and more dignified. He merely exchanged his two filled test tubes for an empty one that Shettles handed him, followed his directions to the men's room on the lower level, collected some of his semen in the tube (using a method decorously known as "manual"), and brought it back up to the foyer, where the doctor was waiting.

"Thanks," Shettles said, his looks and accent making him seem, as an associate once said, like "a larger version of Truman Capote." He hurried off with the test tubes, leaving Del-Zio just a little breathless. "We'll keep you informed."

The name Landrum Brewer Shettles, Norman Mailer once wrote, was like the name Bella Abzug — so resonant that it sounded as if a novelist had made it up. But the doctor was real, and he had spent his long career achieving experimental successes with the temperamental human egg. Shettles was the author of a widely cited photographic atlas, *Ovum Humanum,* that contained some of the first photographs ever made of a fertilized human egg. He was the object of gratitude of hundreds of women, who credited him with bringing joy into their lives with the babies they never thought they'd have. And he was frequently spoken of in the lay press as a visionary, a pioneer, perhaps even a genius.

Yet he was also other things: a thorn in the side of the chairman of his department; a physician whose role at the hospital had been reduced to that of an admitting nurse; an oddball character who kept a collection of clocks all set to the wrong (and different) times; a man who could barely look a listener in the eye. If he was a genius, he was a professionally strangled, socially awkward, personally quirky caricature of one.

Now, moving gingerly, Shettles carried the Del-Zios' test tubes over to the Black Building, whose ornate stonework housed the research laboratories of Columbia Presbyterian Medical Center. This could be the most important moment of his professional life, and he was not about to mess it up by rushing. After months of planning, he was finally holding three vials of precious fluids — two containing extracts from Doris Del-Zio's ovaries, one filled with her husband John's semen — and it was up to him to spin the raw material into gold.

He made his way down West 168th Street, past the hot dog carts and fruit stands, crisscrossing people in long white coats, nurses' uniforms, and the slouchy disarray of medical students. He took the elevator up to the sixteenth floor, to the tissue-culture laboratory of his old friend Mary Parshley, who was retiring. The laboratory was not strictly hers anymore, but she had refused to vacate in July when a younger colleague arrived expecting to take over the space.

Like Parshley, Shettles seemed to have trouble relinquishing territory he thought was rightfully his. He no longer had any official patient care responsibilities or scientific research role at Columbia Presbyterian, yet he was always there, using his hospital ID to gain admittance to places he had no need to be. Dressed in scrubs, his white coat floating out like a reverse shadow, he was seen around the hospital at the oddest hours, seemingly engaged in some interior mission, muttering to himself, so pale and omnipresent that he was known as the Ghost of the Harkness Pavilion. Why was he never at home with his wife and seven children? Where did he change clothes, shower, sleep? And what did he do all night long, without patients to see or a laboratory to call his own?

The skill with which Shettles now manipulated the Del-Zio gametes — the scientific term for the sex cells, egg and sperm — would determine not only the Del-Zio offspring's future, or nonfuture, but his own. Relegated to nurse's work, a pariah in his own department, he was in sore need of some professional restitution.

In the eyes of his hospital superiors Shettles, sixty-four years old and approaching retirement, had never made good on the promise of his early years. When he had joined the medical school faculty of Columbia University more than twenty years before, after completing his residency there, he was the bearer of a prestigious Markle scholarship, a kind of MacArthur "genius award" for biomedicine. Even at that early date, 1954, Shettles was studying human gametes and attempting fertilization in the laboratory. The scholarship, $30,000 to be paid out over five years, was given to just one Columbia scientist a year. It allowed him to travel around Europe visiting obstetrics and gynecology programs in various places, including London, Glasgow, Vienna, and Louvain, Belgium. His goal, he said, was "to try to find out everything I could" about the human egg: "how it grows, how it develops, how it is fertilized."

So much brilliance, so much potential, was conveyed in the shorthand of the Markle scholarship — yet it seemed somehow to have disappeared. Or perhaps not disappeared, really, just twisted into a shape not easily recognizable as brilliance, especially not in a hierarchical institution accustomed to geniuses who came dressed in more conventional garb.

To some, Shettles remained a technological prophet, a single-minded zealot with the intensity and focus of a laser. The department chairman under whom he worked in the late sixties, J. George Moore, described Shettles's research on fertilization as "persevering and visionary." But to others he was the embodiment of a science gone wild with its own sense of infallibility. The gentlest of his critics, such as Howard W. Jones, Jr., of Johns Hopkins University, talked about Shettles's "sheer enthusiasm" and his sincere, almost childish eagerness to jump to conclusions before all the data were in. Those who were less kind called him arrogant, incompetent, or a dangerous combination of the two.

Sheer enthusiasm could probably explain some of Shettles's more intemperate comments, such as those published in November 1973 in a special collection of papers on in vitro fertilization in the *Journal of Reproductive Medicine,* the flagship publication of this emerging subspecialty. In his article, Shettles revealed himself as unwilling to stop to consider the ethical or even the scientific ramifications of his own work. He just wanted to get on with things. "We have invested considerable energies in striving to help infertile couples conceive," he wrote. "Why not continue this mission by constantly advancing our knowledge and skills? . . . Just because the bridge may be out or blocked should not prevent the use of the helicopter."

The helicopter's rotors were whirling even as he wrote those words. "A review of the publications would allow us to project a possible clinical application [of IVF] now, not necessarily in the future," he wrote in the same article. The word "now" should have been underlined. Given the lead time of most scientific publications, it is likely that Shettles wrote those vaguely prophetic words just weeks before his attempt at IVF.

In Parshley's laboratory Shettles got down to business. He pulled out a new, sterile test tube and in it mixed Doris Del-Zio's follicular fluid and tubal mucosa, half a cubic centimeter of sterile human placental blood serum for nourishment and two drops of John Del-Zio's freshly collected sperm. The contents looked a bit murky, but Shettles had no doubt that the test tube contained at least one mature, fertilizable egg and at least one motile sperm ready to do the fertilizing. He sealed the test tube with a black rubber plug, balanced it carefully inside a glass beaker, and placed it in Parshley's incubator. The incubator, which was not designed for human egg preservation, lacked a mechanism to regulate the level of carbon dioxide in its air supply, but it did manage to keep the sample at a toasty 98-plus degrees. The zygote, if there was to be one, needed to be kept at roughly body temperature in order to start dividing.

Then Shettles, confident and excited, stepped out into the hallway, looking for people to tell about the test tube baby he was brewing up on the sixteenth floor.

Shortly after 5:30 that evening Shettles spoke to Dominique Toran-Allerand, a neuroscientist who had arrived at Columbia Presbyterian two months earlier only to find that the laboratory assigned to her was still being used by Parshley. It was quitting time for Allerand, who had been spending her days crammed into one corner of the lab that was supposed to have been hers by now; she was on her way to the parking garage. Early that afternoon she had watched Shettles deposit a test tube in the incubator for what she had assumed was a routine laboratory experiment. Now Allerand learned that he intended actually to implant the culture, after it had grown for about four days, into the uterus of the woman who supplied the egg.

Allerand was horrified. She had seen the gloppy mixture in the test tube, which resembled a frothy chocolate milkshake, and she believed that inserting it into the woman's uterus would lead to infection, pain, and possibly death. She was not a gynecologist or even a

medical doctor — she had a Ph.D. in endocrinology, specializing in the effect on brain functioning of hormones such as estrogen — but she recognized a dirty, dangerous specimen when she saw one.

Before getting into her car to go home, Allerand headed for a pay phone in the Atchley Pavilion and called her supervisor, Georgianna Jagiello, to tell her about Shettles's worrisome plan. Jagiello *was* a medical doctor — her field of expertise was the human ovum, a cell she knew so well that people at Columbia jokingly called her "Mrs. Egg" — and she was worried, too. Allerand's description of the stuff in the test tube made it sound far too turbid to implant safely. At eight o'clock the next morning — it was now Thursday, September 13 — Jagiello called Raymond Vande Wiele, her department chairman, to fill him in on what was taking place in the laboratory down the hall from his office.

Born and raised in Belgium, Vande Wiele was a courtly gynecologist with a list of patients including Faye Dunaway, Greta Garbo, the queen of Morocco, and Sophia Loren. He was refined in all the places where Shettles was raw. An imposing man, he was solidly handsome, with a wide, big-featured face that lit up when he talked about his greatest passion, the New York City Ballet. He loved to play tennis, though his friends said he was a klutz; he also loved travel, reading nonfiction, and wine. When he and his wife, Betty, entertained at their home in Alpine, New Jersey — on a corner of the Frick family estate that they rented cheaply because Vande Wiele was a colleague of one of the Frick sons — he often forced his guests to listen to the latest record he had bought, even before dinner was served. They would sit patiently through, say, the first half of *Tristan und Isolde,* go in for their meal, and then, while the women cleaned up, the men would retire to the living room to hear the rest of the opera.

Vande Wiele did not especially care for Shettles; in fact, his files were filled with carbon copies of the hostile letters he had written to Shettles, trying without success to rein him in. Shettles dated their animosity to the time he confronted Vande Wiele with what he con-

sidered the ugly truth of his family history: that Vande Wiele's older brother, Jos, had collaborated with the Nazis occupying Belgium during World War II and had served as an officer in the SS.

Vande Wiele had a hot temper, but he didn't express it often. Indeed, he didn't even express it to Jagiello when she phoned him about the strange vial in Parshley's incubator, though he later recalled being profoundly enraged that day. After speaking with Jagiello, Vande Wiele called his two supervisors, the acting dean of the medical school and the hospital president. Then he phoned Allerand and told her to go into the laboratory she shared with Parshley, open the incubator door, and remove the test tube she had watched Shettles place there the previous afternoon. Bring it to me, he instructed.

And so it was that by nine o'clock that morning, Raymond Vande Wiele had the Del-Zio test tube sitting on a coffee table in his office, held upright in the same beaker that had propped it up in the incubator. Now he had to decide what to do with it. Take out the stopper and contaminate it so no one would dare implant it into anybody's uterus? Toss it out completely? Leave it out at room temperature, effectively stopping the cell division? Or return it to the incubator and allow the experiment to proceed?

2

The Dance of Love

> Working forgetful of his family and regardless of his friends, he bent solitary to subtle tasks in still nights.
>
> — Paul de Kruif, *Microbe Hunters* (1926)

William pancoast, one of the most esteemed physicians in nineteenth-century Philadelphia, was a solid, honorable, devout man. After serving with distinction in the Civil War, he took a job at the Jefferson Medical College and followed in his father's footsteps to become a professor of descriptive and surgical anatomy there in 1873. Yet in 1884 Pancoast made a medical decision that, by today's standards, reflected the height of medical arrogance — and that eventually became the first documented case of the creation of a "test tube baby."

It began when a wealthy young Philadelphia couple sought out Pancoast, hoping he could do something about their continuing inability to have a child. In those days doctors tended to blame any fertility problem on the woman; there was a long, sorry history of childless couples divorcing so the husband could move on from his "barren" wife and try again with somebody supposedly better equipped. Pancoast began in the usual way by examining the wife. When he found nothing wrong with her, he did something surprising: he examined the husband. The young man admitted to a dalliance that had led to a bout of gonorrhea; could that be relevant? Yes, it could, since gonorrhea often destroys a man's ability to make sperm. Pancoast tried to treat the gonorrhea, which in those days meant infecting the patient with malaria in hopes of inducing a fever that would kill the gonorrhea germ. But that didn't work; the wife still did not become pregnant. Then one of Pancoast's medical students had an idea. Since venereal disease had clearly destroyed this man's sperm, why not use sperm from some other man?

Pancoast agreed to the plan. One day, when he was ostensibly examining the wife, he went into another room, where one of his "most handsome" medical students was waiting by prior arrangement. The young man had collected a sperm sample, which he handed to Pancoast, who took it back into the examining room. The wife had been anesthetized; she was unconscious when Pancoast drew up a bit of the fluid in a rubber syringe, inserted the syringe into her vagina, and, when the tip reached her cervix, gave a good forceful squeeze.

Nine months later the delighted couple became the parents of a beautiful baby boy. Much to Pancoast's relief, the baby looked like the husband.

In time Pancoast began to regret having acted in secrecy — though he never regretted his action. After a few years, with great trepidation, he took the husband into his confidence. If the parents knew the truth, would the wife reject the child that had grown in her womb and was now pulling at her heartstrings? Would the husband resent it as a bastard, evidence of a medically induced cuckolding? What a blessing to find that the husband was unruffled by the news. But he made Pancoast promise never to tell his wife.

When one of Pancoast's medical students wrote an account of this event twenty-five years later, the public reacted with indignation. Many people, physicians and laymen alike, thought such "ethereal copulation" was horrible, dangerous, too great an intrusion into the natural order of things. Letters to the editor of *Medical World,* in which the account was published in 1909, said Pancoast had in effect raped his patient under anesthesia, and his actions would have been less objectionable had the medical student just had sex with her.

Over the next few decades artificial insemination by donor (AID) became more common, but it continued to raise troubling questions. In 1936, for instance, *Literary Digest* took what it called a "sober, scientific" look at AID. But it unintentionally reflected the era's squirmy notions about the procedure and about the meaning of a genetic connection between parent and child. A sperm donor should be of the "same race stock" as the husband in the infertile couple, wrote the author, to "avoid the tension which might develop later when [for example] the growing child, if of mercurial Italian heredity, might clash with a legal father of phlegmatic Germanic stock." And physical inconsistencies could prove just as much a threat to family equanimity: "if the father and mother both are brown-eyed, but the child has blue eyes, a certain psychological irritation might develop." (A strange example to choose, since two brown-eyed parents who both carry a recessive gene for blue eyes can have a biological blue-eyed child the old-fashioned way.)

As late as the 1950s, some people believed that AID should be forbidden, lest it metamorphose into something more unnatural and more dangerous. The slippery slope argument again, this time one that flings us down to a society where alien changelings take their places in otherwise ordinary families. In many states artificial insemination with another man's sperm — even if the husband gave his signed approval — was tantamount to adultery and could be grounds for divorce. Worse, a child born of that act was considered illegitimate. A 1954 case in Illinois, *Doornbos v. Doornbos*, concluded that a "test tube baby" conceived through AID, even if the procedure was undertaken with the husband's consent, was legally a bastard.

A similar sentiment permeated Great Britain. In 1959 a special committee of the Eugenics Society produced the Feversham Report, which concluded that artificial insemination by donor was ethically and morally repugnant. The report stopped just short of recommending that the procedure be made a criminal offense, as some groups had proposed. But it raised a jumble of objections to its use, many of which would arise a few years later in response to IVF: the risk of chromosomally damaged babies, the derangement of ordinary family relationships, the incursion of science into matters of procreation best left to nature and to God.

AID turned out to be a slippery slope with a mild gradient and a soft landing. As the social opprobrium of adultery and illegitimacy eased, so did the most powerful objections to artificial insemination. By the second half of the twentieth century, AID was considered a private family matter, and the procedure was responsible for the births of some 20,000 American babies every year. In spite of *Doornbos v. Doornbos*, most states recognized the man who raised the child as its legal father, even though his genes had not helped create the baby.

Just as AID was becoming generally accepted, there emerged a new, more intrusive form of reproductive technology, which involved manipulating eggs as well as sperm. And this technique was far from acceptable; indeed, it looked as though it might never become generally tolerated, much less embraced. This was test tube baby making in its more modern meaning, in vitro fertilization.

The first successful lab manipulation of mammalian eggs, in 1934, resulted in a litter of seven dark gray rabbits. The babies were born to a New Zealand Red doe, but their biological mother was an agouti female whose eggs had been mixed with semen from a black male in the Harvard laboratory of Gregory Pincus. Pincus mixed the gametes together in a watch glass (the small scoop of crystal that protects the face of a pocket watch), then immediately transferred them into the agouti's oviduct, where the actual fertilization occurred. This wasn't in vitro fertilization in the strictest sense because the egg was fertilized in vivo (inside the animal), but it was the first indication that IVF was possible in mammals and, by extension, in humans. A few years after Pincus's accomplishment, an unsigned essay in the venerable *New England Journal of Medicine* discussed this line of research using uncharacteristically jaunty language. The author of "Conception in a Watch Glass" wrote that "if such an accomplishment with rabbits were to be duplicated in human beings, we should, in the words of 'flaming youth,' be 'going places.'" Assisted reproduction would be a wonderful development: "What a boon for the barren woman with closed tubes!" Soon "another link may be welded in the chain by which mankind strives to hold nature under control." Judging from the language and the enthusiasm, people in the know thought that the anonymous author was one of Pincus's colleagues at Harvard, the eminent gynecologist John Rock.

In 1944 John Rock and a colleague from Harvard, Miriam Menkin, took the first tentative steps toward human IVF. Rock and Menkin retrieved a woman's mature eggs during abdominal surgery for some other condition, washed them in a laboratory preparation called Locke's solution, bathed them in serum taken from the woman who had donated the eggs, and incubated them for twenty-seven hours. Then they took a sample of sperm, which they washed and suspended in Locke's solution, and mixed the gametes together in a petri dish. When their report of the first successful fertilization of a human egg in vitro was published in *Science*, scientists were skeptical. Some said that even though Rock and Menkin had let the fertilized egg grow for a while, they had not tried to transfer it into a uterus, making it unclear that the egg had really been fertilized in the first place. Maybe, critics said, the observed cleavage was just a laboratory artifact, something that can happen to unfertilized eggs in a petri dish — a kind of reflex cell division that will never lead to a healthy, normal child.

After Rock and Menkin's report, some scientists wanted to go no further; they feared what would happen if they learned how to mix up life in the laboratory. But others were driven to accomplish this same feat — driven by several impulses, some of them contradictory, often coexisting in the same individual. At work, and at odds, were the desires to serve the needs of mankind, to capitalize on a new market, to discover something for the pure joy of the discovering, and to win the honor and fame of being first. And all of these motivations came up against the competing impulse to maintain the status quo, which could be found at the heart of the views expressed by many groups acting as the keepers of cultural norms: organized religion, political conservatives, and influential private citizens who longed for the good old days when everything was supposedly simpler.

It was a delicate balance, this walk along the tightrope between too much progress and too much restraint. The balance was put to the test in 1961, when an Italian scientist made a startling announcement. Daniele Petrucci, an embryologist at the University of Bologna, said he had created forty human embryos through IVF and, even more shocking, that he had grown one of them in the laboratory for nearly a month. That one embryo, he said, had developed to a point where it had a discernible heartbeat. But ultimately it became, as one newspaper account put it, "deformed and enlarged — a monstrosity," and Petrucci destroyed it after twenty-nine days.

The Catholic Church was horrified. Pope Pius XII had already gone on record as opposing any sort of external human fertilization, even of the type that Rock and Menkin had performed at Harvard. Such scientists, he said, "take the Lord's work into their own hands." But Petrucci's work went a significant step beyond. *L'Osservatore Romano,* the Vatican's official newspaper, blasted the experiment as "monstrous" and "sacrilegious" — especially when the scientist was an Italian who described himself as an observant Catholic. The only acceptable way for life to begin, wrote the Vatican, was with "the most supreme assistances of love, nature and conscience."

Scientists, for their part, were skeptical that Petrucci had even done what he claimed, though growing an embryo in the lab was a scientifically credible achievement at the time. It had already been done with mouse embryos at Cambridge's Strangeways Laboratory, and the researchers had kept the embryos alive for a comparable amount of time, one-tenth the length of a normal gestation. But it was the way Petrucci announced the news that raised suspicion: reporting it in the popular press — the Roman newspaper *Paese Sera* — without any follow-up article in a reputable scientific journal. How reliable could such a claim really be? A few months later Petrucci announced in another newspaper that he had managed to keep a human embryo alive for forty-nine days. After seven weeks of growth in the lab, he said, this second embryo died because of what he called a "technical mistake."

His work repudiated in his home country, as well as in much of Western Europe and the United States, Petrucci went on to become something of a hero in the Soviet Union. He spent two frigid winters at the Institute of Experimental Biology in Moscow, teaching the Soviets how to keep embryos alive in artificial wombs. Soon the Moscow scientists claimed that they had cultivated 250 human embryos for staggering lengths of time — one fetus for a full six months, two-thirds of the way through a normal pregnancy. Before it died, they said, it weighed one pound two ounces.

Robert Edwards was conducting work along the same lines as Petrucci, but he was never interested in growing embryos outside the body. His mission was to allow barren couples to have babies through normal pregnancies in normal human uteri, not through cultivation in some artificial womb. Wrapped up in that mission were other, less altruistic motives — the same motives at work in any creative enterprise in which the stakes are high — but at its core was Edwards's desire to help people have their own biological babies.

Edwards worked in the physiology department of Cambridge University in England, just across Downing Street from the famed Cavendish Laboratory, where James Watson and Francis Crick had discovered the double-helix structure of DNA in 1953. Edwards's calling revealed itself to him one day in 1960 in a sparkle of insight, much the way the double helix had uncoiled in the imaginations of Watson and Crick. He and his wife, Ruth, who was a scientist herself and the granddaughter of the physicist Ernest Rutherford, had two daughters under the age of two in the fall of 1960, and a childless couple in the neighborhood often dropped by to peek inside the pram. "When they visited us that autumn and cuddled our babies," Edwards later wrote, "I could not but be aware of the feelings aroused in them. The trees bore fruit, the clouds carried rain, and our friends, forever childless, played with our Caroline, our Jennifer."

So he decided to find a way to help "forever childless" couples like his neighbors. But to accomplish this, he needed a reliable source of research material. He had been studying mouse eggs, but he knew he had to move on to human eggs if he was to affect human fertility. He appealed to his university colleagues, to the gynecologists at the local hospital, to scientists and physicians in London and even beyond, asking to beg or borrow whatever spare eggs they could retrieve from patients after gynecological surgery. Generally he met with wariness, an unwillingness to get involved in so controversial an undertaking. Only one physician offered enthusiastic assistance: Molly Rose, an obstetrician-gynecologist at Edgware General Hospital in north London, who had delivered the Edwardses' first daughter, Caroline, in 1959. Rose agreed to send him ovarian tissue whenever she performed operations for polycystic ovarian syndrome, a hormonal disorder that can cause infertility, obesity, excessive hairiness, menstrual irregularities, and enlarged ovaries. Although the syndrome is associated with the release of an abnormal amount of gonadotropin from the pituitary gland, gynecologists at the time were treating the problem by operating on the ovary, not the pituitary. They would perform a "wedge resection," cutting away a long piece of ovary like a slice of watermelon. Some of those wedges contained immature eggs, known as oocytes, and Edwards eagerly accepted them.

But the wedges presented problems of their own. Oocytes, the precursors of fully mature eggs, or ova, are plentiful in ovarian tissue and are relatively easy to retrieve. But they are especially hard to fertilize, primarily because they have to be brought to maturity first. In the 1950s at Harvard, embryologist Min-Cheuh Chang, working with rabbits, discovered how to turn mammalian oocytes into ova. But whenever he tried to impregnate a rabbit using lab-matured ova, something went wrong. No matter what he did, embryos produced from eggs that had matured in the laboratory rather than in the body never lived in the uterus beyond a couple of weeks. It was almost as if Nature was fighting to protect her privilege; whenever Man made an apparent breakthrough, Nature formed a new line of defense. Would the same mysterious malformation interfere with IVF attempts involving human eggs matured in the lab?

This question haunted Edwards, as did another: would IVF increase the likelihood that more than one sperm could fertilize an egg? Normally thousands of sperm compete for entry into the zona pellucida (literally, "translucent region"), a sort of halo around the egg that serves as a gummy protective coat. When one sperm burrows through, an instantaneous chemical change makes the zona pellucida impermeable to any other sperm. This is a critical step; if two sperm were to get through, the resulting zygote would have extra chromosomes, making it so grossly abnormal that it would die within days. Edwards could not know for sure whether fertilization in a petri dish would interfere with the natural sperm-blocking mechanism.

Even if only one sperm did the fertilizing, it could easily be a damaged or defective sperm. In natural conception, only the strongest sperm — a few thousand out of the 60 million or so in a typical man's ejaculate — manage to swim far enough up the fallopian tube to reach the egg. This initial race eliminates all but the hardiest sperm. In vivo (that is, in the body instead of in a petri dish), scientists have found, human sperm that reach the fallopian tube are better able to penetrate the zona pellucida than sperm that make it only as far as the uterus. In addition, something in the female reproductive tract, possibly at the junction between the uterus and the tube, seems to serve as a physical barrier to abnormally shaped, and possibly abnormally functioning, sperm. Because IVF short-circuits both of these safeguards, weak sperm have a better chance of fertilizing the egg than they would in nature, possibly undermining the vigor of the next generation.

These questions, which bothered Edwards, didn't really faze the more single-minded researchers such as Petrucci in Italy and Shettles in New York. Shettles was almost fanatically devoted to fertilizing and culturing human eggs. Working alone in his small laboratory, he conducted experiments on the few immature oocytes that he managed to collect from patients undergoing gynecological surgery. Usually, following the convention at the time, Shettles would take a patient's eggs without telling her. Most researchers, including Shettles, saw the removal of ova as just a matter of taking a little extra tissue during an operation that was being done anyway for some other reason. Since it added no risk to the procedure, what could be the harm? Shettles collected oocytes (he did this so often that he was once called, in a magazine article, an "egg poacher") and matured them in a culture medium pieced together from bits of the egg donor's ovarian follicle, plus scrapings from the walls of her fallopian tubes. A few days later, when the oocytes had matured into ova, he exposed them to sperm — sperm that often came from Shettles himself.

He would peer through a microscope at a lone egg in a petri dish, surrounded by sperm trying to invade the egg's sticky outer layer. There the egg would sit, large and inert, like some microscopic Buddha, as it was pounded by hundreds and hundreds of sperm competing for the chance to be the single one to penetrate to the mother lode. And Shettles would patiently stare — taking photographs all the while — at the cellular frenzy he liked to call "the dance of love."

He watched this love dance again and again, growing what he later said were "several" test tube embryos to the morula stage before destroying them. Except once, in 1962, when he supposedly went one step further, transplanting the morula into a woman's womb. According to Shettles, the transplantation was successful. If this claim was true — a claim that many experts in the field found reason to doubt — Shettles, then fifty-three, was the first man in the world to create a pregnancy with a lab-fertilized human egg.

The 1962 transplantation began just like all the other fertilizations. As Shettles later described it, he removed an immature egg from a woman who was undergoing surgery on her fallopian tubes, matured the egg in a petri dish, fertilized it with sperm from her husband, and let it grow in the lab for five days until it reached the blastocyst stage. Then, he said, he found another woman who was scheduled to have a hysterectomy — surgical removal of the uterus — to treat her cervical cancer. A few days before the hysterectomy, while the first woman's embryo was still growing in its laboratory dish, Shettles said he explained to the second woman what he wanted to do and asked permission to introduce the lab-grown embryo into her uterus. No strings attached, he told her; even if the embryo took up residence in her uterus, it would be out of her body, along with the womb itself, in a few days.

This was years before the concept of informed consent even existed; it was rare in the early sixties for a scientist to lay out to a patient the goals of his experimentation and its risks and benefits in a clear and structured way. With the standards of experimental ethics set so low, Shettles may have told his prospective IVF subject a lot or he may have told her very little. Maybe she did not really understand what he wanted to do or realize that implantation would be a groundbreaking achievement if it worked. Or maybe she did understand, viscerally as well as intellectually, because she herself had gone through the agony of infertility; she was thirty-six years old and had already had both of her fallopian tubes removed. Maybe she just wanted to do her part, at very little risk, to help other infertile women have the chance to have their dreams come true.

The woman agreed to the procedure, and two days before her hysterectomy, Shettles said, he introduced the embryo into her uterus in an office procedure using a pipette and catheter inserted into her vagina. When her uterus was removed two days later, he opened it up — like a hamburger bun, he said — and saw that the embryo had indeed implanted in the lining. Over the course of two days it had grown to the size of several hundred cells, and it appeared to be normal. He said that he saw no reason why the pregnancy would not have continued had it been allowed to progress.

Shettles's accomplishment was, if true, nothing short of astonishing. Working alone in a patchwork lab, keeping his specimens in the same hallway refrigerator where his coworkers kept their lunch, he was claiming to have succeeded in a technique that had foiled some of the leading contenders in the race to create the first test tube baby, scientists with much more financial backing than he. And he was claiming to have done it before anyone else in the world.

So why didn't he announce his feat with bugles blaring — or at least, in the time-honored medium of scientific crowing, through publication in a peer-reviewed journal? The race was sure to be winner-take-all, and it looked as though Shettles had pulled out in front. The stakes were high: an international spotlight and a place in the history books. "The publicity that would be attached to the first success in producing a 'test tube baby,'" wrote one observer, "would almost certainly push a mere heart transplantation or moon walk off the front page of any national daily." It would be surprising indeed if Shettles weren't interested in at least a piece of that publicity.

Fame is, admittedly, a dubious motivator for scientific achievement, especially since front-page appearances often get scientists in trouble with their professional peers. But the longing for fame is insistent, even for research scientists, who are supposed to be above such things. Like their counterparts in every other type of contest, they want to be first; there is no glory in being second.

When South African surgeon Christiaan Barnard performed the world's first heart transplant, he became an overnight sensation. It didn't hurt that he had the square-jawed, lean-limbed ease of Steve McQueen and a reputation as an international playboy, but he did not become famous for his looks or his love life. He became famous because he had broken a barrier.

He had also established a market niche: a population of potential patients who never would have imagined that one way to repair a damaged heart was to exchange it for a new one. And if a heart could be transplanted, other organs could, too. By 1967 some kidney transplants were being done, and some universities were trying to distinguish themselves as transplant centers. Enterprising surgeons could see the potential for transplantation of livers, corneas, bone marrow, and all the other organs and tissues that can be damaged in the course of a lifetime. Their efforts revealed that market forces can be a powerful scientific stimulus. As long as patients were willing to pay, physicians would line up to find new ways to fill a medical need — or, in some cases, to create one.

Were nobler, more altruistic impulses involved as well? Yes, of course; helping patients has always been an important motivator in biomedical research. Transplant surgeons most certainly wanted to prolong the lives of their mortally ill patients, just as scientists working on in vitro fertilization wanted to help their distraught patients who were longing for babies. Robert Edwards, who was thirty-seven in 1962 and one of Shettles's most formidable competitors in the IVF race, used to say that his goal was nothing so grand as "creating life"; his goal was simply to improve it. "Life is uniquely *there*, bursting out everywhere, a wonderful part of our universe," he said. "The most that any of us can do is to help to make life possible, and to make it healthy and good. And that [is] what I [want] to do — to help to make human life possible, and healthy, and good."

Other, less extreme techniques could also deliver babies into patients' aching arms; other methods could make life "possible, and healthy, and good." But only IVF was a scientific leap worthy of a hero's spirit. Only IVF could get Edwards some of the other goodies that he — like Shettles, like just about any scientist who was honest about it — was after: fame and money; the respect of his peers; a shot at the Nobel Prize; funding for his research; and, perhaps most compelling, the simple thrill of winning.

Given the intensity of the race, then, why didn't Shettles stake his claim for a successful embryo transplant in 1962, as soon as it occurred? In truth, there were so many uncertainties about the details of Shettles's implantation experiment and so few records — no hospital or laboratory reports, no original slides, no accurate account, even, of whether it took place in 1962, or 1961, or 1964 — that writing an article about it would not have been easy, and getting it through peer review would have been more difficult still. The lack of specifics made some observers wonder whether the embryo transplant had actually happened, as did the stunning coincidence of the embryo's development and the uterus's readiness being, merely by chance, in such perfect hormonal synchrony.

Whether or not Shettles had successfully transplanted a human embryo, one thing was clear: Edwards was having no luck. Part of his trouble came from having very few ova to work with, even with Molly Rose's cooperation. If he had access to more eggs, he reasoned, he could accumulate more experience and increase his chances for success. Maybe he would have better luck finding eggs in the United States.

Edwards turned to America at the height of its love affair with science. The New York World's Fair, which opened in 1964, took as its theme a reverence for the power, drama, and unmitigated good of scientific progress. The gleeful hoofers in the show at the Du Pont pavilion, "The Wonderful World of Chemistry," captured the spirit of the entire fair. Built on 640 acres of filled swampland in Queens, the World's Fair forecast a rosy Tomorrowland that would blossom in four dimensions: the Information Age, with computers putting people in touch with data instantaneously; the Space Age, when America's race with Soviet cosmonauts would put a man on the moon; the Consumer Age, bringing modern conveniences that would make ordinary life a breeze; and the Atomic Age, with nuclear power providing a source of electricity "too cheap to meter."

In this can-do environment, it wasn't surprising that when Edwards wrote to his old friend Victor McKusick in the fall of 1964, hoping to find one kindred soul among the thousands of scientists in America, he got a positive response. McKusick, a geneticist at Johns Hopkins University in Baltimore, put Edwards in touch with not just one kindred soul but two: Howard Jones, a gynecological surgeon, and his wife, Georgeanna, a reproductive endocrinologist, both of whom were right there at Johns Hopkins.

Edwards arrived in the States on a research visit in July 1965, leaving behind a family that by now numbered five daughters under the age of six (the last two were twins). On his first day at Johns Hopkins, he was taken over to where the Joneses were eating in the cafeteria. Howard Jones had a hard time keeping a straight face when he met Edwards; cursed with a sense of humor that got him into trouble whenever a private joke sent him into giggling fits, Jones kept remembering that when McKusick had first phoned to see if he would meet a "mouse geneticist" from England, Jones had conjured up "an image of a mouse who knew a lot about genetics."

With his shaggy hair, big spectacles, and earnest brown eyes, Edwards looked more like a cocker spaniel than a mouse. He sat down across from the Joneses and started in on his sales pitch. In a rapid-fire whisper, he told them of his plan to fertilize human eggs in vitro as a way to circumvent blocked fallopian tubes, of his success in fertilizing mouse eggs, and of his need to do as many human fertilizations as he could if he was ever to help women suffering from infertility.

Then he paused. It might have been difficult at first to recognize it as a pause; Edwards often punctuated his staccato speech with long hesitations, partly for emphasis, partly as a kind of silent stammer. Whenever he gave this recitation to colleagues in Britain, it was at this point that his listeners would start to shuffle and begin making excuses about having to be somewhere else. But the Joneses were different.

"We'll do what we can to help," said Howard, a slim man with a prominent nose, whose sonorous voice and white hair conveyed far more seriousness than did the twinkle in his eye. From that point on, the three were bound together on a scientific odyssey that would, within a few short years, turn them into heroes, pariahs, lightning rods, punching bags — a whole range of metaphorical social roles, from adored to reviled, for which they were woefully unprepared.

Edwards's arrangement with Howard Jones was much like the one he had with Molly Rose: each time Jones did a wedge resection for polycystic ovarian syndrome, he would hand over the extra pieces of ovary to Edwards. For the first three weeks Edwards used these precious human oocytes to study meiosis, the process, little understood at the time, by which an egg reduces its chromosomes by half. (The sperm undergoes meiosis, too, so that after the egg and sperm unite in fertilization, the resulting zygote has the normal number of chromosomes, rather than twice the normal number, as it would otherwise.) Satisfied that he now understood meiosis, Edwards tried again and again in the following three weeks to fertilize the matured eggs in a petri dish, using the most readily available fertile sperm he had: his own.

One of Edwards's goals that summer was to figure out the process known as sperm capacitation. In nature, sperm are not ready to penetrate the egg's zona pellucida until after they have traveled through the female reproductive tract. Knowing this, scientists believed that something in a woman's vagina, cervix, or fallopian tube gave a signal to the sperm to turn on the proteins that allowed it to penetrate the shimmery barrier of the zona. They just didn't know what it could be.

Capacitation drove Edwards all summer. He mixed bits of uterine tissue, tubal mucosa, and vaginal secretions into his culture dishes, hoping that something in one of these potions would switch on, or capacitate, the sperm. He collected sperm from the vaginal tract of some of Howard Jones's patients, who had been instructed to come into the laboratory immediately after having intercourse. He inserted human eggs and human sperm into the fallopian tubes of rhesus monkeys, hoping there was some magic in the tubes, even monkey tubes, that would enable the human sex cells to fertilize.

When he wasn't working in the laboratory, Edwards socialized with the Joneses, who were the only people other than McKusick whom he knew. As a couple the Joneses were perfectly matched: similar height, similar hair and facial expressions, similar passions. Georgeanna Jones was not conventionally pretty, but her face was warm and animated and beautiful in its way, with an openness that made people eager to confide in her. Edwards took to her easily, and to the mischievous Howard as well. He often accompanied the Joneses and their three children out on Chesapeake Bay, where they loved to sail. And the Joneses taught him the intricacies of cutting through the hard, spicy shell of a steamed Maryland crab with mallet and pliers in a near surgical approach to getting at the sweetest meat.

The Joneses, as Edwards could see just by looking, had a kind of fairy-tale marriage, a meeting of two intensely individual people who fit each other in all the most interesting places. They had known each other their whole lives — Georgeanna's father, an obstetrician in Baltimore, had delivered Howard in 1910 — and managed to share their related careers without any of the rivalries that bedevil similar marriages. In a way, Georgeanna, who had had a lifelong interest in reproduction, primarily as an endocrinologist, was the intellectually stronger one; Howard had changed his specialty from cancer surgery to infertility surgery so his career would better complement hers. But for all her professional fierceness, Georgeanna was a traditionalist at home. If any of the three kids needed a parent, it was Georgeanna who was usually on call. And she insisted, when the children were small, that their friends call her "Mrs. Jones," never "Dr. Jones." "Dr. Jones" was Dad, not Mom.

Despite his frustrations with capacitation, Edwards forged ahead at Hopkins. By some measures, he seemed to be succeeding. Working alongside Howard Jones, he exposed the eggs to sperm and then was able to spot two pronuclei, the genetic packages produced at meiosis that contain just twenty-three chromosomes each. One pronucleus comes from the egg, one from the sperm, and after fertilization they gradually drift together to form a normal, forty-six-chromosome nucleus. Finding two pronuclei in a single egg was often taken as a sign of successful fertilization.

But it wasn't proof: two pronuclei occasionally meant that there had been a problem during meiosis and that both of the pronuclei belonged to the egg. This would be the first step in the rare process known as parthenogenesis (literally, "virgin birth"), in which an egg that has mistakenly held on to its full complement of genes starts to divide into two, four, eight cells the way a developing embryo usually does, even though a sperm has never penetrated. Because of the possibility of parthenogenesis, the definitive sign of fertilization was not two pronuclei, or even a cleaving zygote, but something much simpler: the sighting of a sperm tail inside the egg. Edwards was unable to find a single tail in any of his laboratory cultures. He left Baltimore believing that he and Jones had failed to achieve fertilization.

In the autumn of 1965 Edwards was back in England with his wife and daughters and back to the continuing scientific isolation of Cambridge. Molly Rose kept supplying him with eggs, but that wasn't what he needed anymore; he knew how to work with human eggs in the lab. And, he soon realized, he was even able to fertilize them. At home in his familiar lab, he had an epiphany about the chemical he and Jones had used to prepare slides in order to take photographs under the microscope to document their results. He realized that the chemical was destroying the sperm tails. The sperm had gotten inside the ovum — the second pronucleus was evidence of that — but the tail, the footprint of its presence, was gone. Edwards and Jones had, without knowing it, fertilized human eggs outside the womb in the summer of 1965. This was a landmark accomplishment, the first such in vitro fertilization in the world. That is, until Landrum Shettles claimed, several years after the fact, that he had beaten Edwards and Jones by a full three years.

Sperm capacitation, which had seemed so crucial, turned out to be irrelevant. None of those bits and pieces of femaleness that Edwards and Jones had been tossing into their laboratory stew turned out to make a difference. Once he understood that the fertilizations had been successful, Edwards realized also that human sperm became capacitated in the laboratory simply by being washed.

What Edwards needed now was a chance to transfer his fertilized eggs back into a woman's womb — the first step in helping couples have babies of their own. And for that he needed some infertile patients.

Patrick Steptoe had plenty of infertile patients. A patrician-looking man with thinning silver hair and thick horn-rimmed glasses, Steptoe had a busy gynecological practice in Oldham, Lancashire. The position was not the kind of prestigious research appointment he had once dreamed of, but it was good enough. At first, the move to the town, not far from industrial Manchester, had felt like exile to Steptoe, his wife, Sheena, and their two young children. But within a few years the Steptoes discovered that many of the pleasures they had come to love in London — theater, concerts, cricket matches, sailing — were available in their new environs as well.

Less easy for Steptoe to adjust to was the nature of his practice. From a young age, he had wanted to help infertile couples have children, but in Oldham he was kept busy dealing with the medical problems of a town that had been too long without a local gynecologist. In the context of pelvic tumors grown huge and prolapsed uteri gone unrepaired, Steptoe had trouble finding time to help otherwise healthy young couples have babies.

Finally, after five years of whittling away at the backlog, Steptoe saw his workload ease, and he was able to devote himself to some of the patients who most intrigued him. One technique he worked hard to develop was a new way to view a woman's internal reproductive tract. It offered a much clearer, more accurate view than the techniques then available — curettage, x-ray, tubal insufflation, culdoscopy — and was much less invasive than laparotomy, which required general anesthesia and opening up the abdomen to get a good, clear view of the organs inside.

The new technique was laparoscopy, which involved inserting into the abdomen a slim, flexible fiberoptic tube through which a surgeon could peer, periscope style, at the ovaries, uterus, and fallopian tubes. And in the right hands, it could also be used to do something that had been impossible with any other instrument: retrieve fully mature human ova. Steptoe suspected that he just might have the right hands.

The Albert and Mary Lasker Foundation's annual awards have been called, informally, the American Nobels, since so many recipients have gone on to receive Nobel Prizes. Each year the foundation hosts an elegant luncheon in Manhattan and announces prizes for that year's most notable contributions to clinical medicine, basic science, and public service. In 1966 the award for public service went to Eunice Kennedy Shriver for her efforts in mental retardation. What really interested Shriver at the time, however, were the social and ethical implications of biomedical technology. Every new development, she knew, implied some sort of moral choice, and she worried that the wrong people were being allowed to make these choices. She worried, essentially, about the slippery slope.

The Lasker Awards luncheon on November 17, 1966, was held at the Plaza Hotel on Fifth Avenue. Shriver's mother, Rose Kennedy, her brother Robert, and her husband, Sargent Shriver, director of the federal Office of Economic Opportunity, all came to New York to watch her receive her honor. In her acceptance speech, Shriver offered an unusual twist on the slippery slope idea: a reminder that even beneficial developments can be the first step on a slope that might lead to great harm.

Because of lessons learned by behavioral scientists, Shriver said, a profoundly retarded individual — or as she put it, using the terminology of the time, a "helpless retardate" — could now be taught to feed himself, dress himself, and talk. This was, obviously, a good thing. But "if the knowledge obtained through experimental psychology can accomplish such wonders today," she asked her audience, "will not the future see other ways of controlling human behavior? Who will decide how behavior shall be controlled, for what purposes, and according to what standards?"

Behavior control was only the beginning, she said, and much could go wrong if the wrong people made important decisions about it. "Someone lacking in moral judgment will surely advocate euthanasia as a solution to the population problem," she said. Or a similarly amoral person might suggest that it would be "more efficient and less expensive to do away with the people over sixty-five who are ill and tired of living." Or maybe the government would create a board to decide who should be allowed to reproduce, on the basis of their genetic vigor. "A whole new profession could be developed," she said, "like race track prognosticators, who would predict exactly which parents would produce the best progeny for success."

Clearly, science was progressing in ways that required some sort of oversight. The government had already made a move in that direction: that same year the National Institutes of Health (NIH) had adopted a formal set of ethical standards for all research conducted under its aegis. But no mechanism had yet been devised to assure that NIH grantees were sticking to those standards . And it was not quite the kind of oversight that Shriver was advocating. She envisioned something more formal, removed from the old-boy network of biomedical researchers. Scientific progress, according to Shriver, was too dangerous to be monitored by the scientists themselves, whose motivations might color their judgment and might not reflect the beliefs of society at large. It was important, she said, to choose the right people, those trained in biomedical ethics rather than biomedical research — a tall order, since at the time not one American university had a formal graduate program in bioethics. She said it would take men and women trained in such fields as religion, law, and moral philosophy to ask the truly profound questions: questions about good versus evil, ethics versus expedience, life versus death.

Shriver closed her remarks with a quote from Pierre Teilhard de Chardin, the French priest, paleontologist, and philosopher who spent his life trying to reconcile science and religion. She turned to Teilhard de Chardin to make the point that science must be carefully controlled, yet the quote she chose idealized the mythic figure of Prometheus — the quintessence of unfettered scientific progress for its own sake: "Someday, after mastering the winds, the waves, the tides and gravity, we shall harness for God the energies of love. And then, for the second time in the history of the world, Man will have discovered Fire."

One short year later, biomedical technology leapt into the territory of mixed blessings that Shriver had warned about. As Christmas 1967 approached, newspapers blazed with reports that Christiaan Barnard, a charismatic forty-five-year-old cardiac surgeon from Cape Town, South Africa, had accomplished what was, depending on one's point of view, either a medical miracle or an act worthy of Dr. Frankenstein. He had implanted the beating heart of one person into the chest of another. The patient, Louis Washkansky, was a grocer in his mid-fifties suffering from diabetes and heart failure. The donor, twenty-five-year-old Denise Darvall, had been struck by a car and had suffered irreversible brain damage. "Jesus," Barnard said in Afrikaans when Darvall's heart started beating in Washkansky's chest, *"dit gaan werk"* — it's going to work. A few days after the operation, Washkansky told his nurse, "Look at me — I'm the new Frankenstein." But it was Barnard, of course, who was the modern analogue of Dr. Frankenstein — and his first creation died of pneumonia after eighteen days.

Like IVF, heart transplantation led to a reexamination of the borders between life and death, between God's work and man's. Even though other organs, such as kidneys, had already been transplanted without much social comment, the heart was something else: personal, intimate, the romantic's wellspring of kindness and love, so loaded with significance that you might wonder who you really were if you were fueled by someone else's. The *Johannesburg Sunday Express* had a fine time riffing on the many connotations of the heart: the "tender heart, the aching heart, the bleeding heart, the broken heart. The hard heart, the soft heart; the heart that sends its golden tendrils to the furthest corners of the body and the heart which reputedly houses the dark chambers of the soul." The heartbeat is one of the elemental signs of life (breathing being the other) — how, then, could society tolerate surgeons removing a donor's heart while it was still beating? How could surgeons have the audacity to capture that beating heart, to take it out of one chest, hold it pulsing in their hands, and place it in the chest of a stranger?

The technological development of heart transplantation forced a new definition of something that had previously been taken for granted: death itself. In 1968, stimulated in part by Barnard's revolutionary operations, an ad hoc committee at Harvard Medical School issued a new definition of death as "irreversible coma" or "brain death." As laid out in the so-called Harvard Standards, brain death was different from the more traditional measures, absence of heartbeat or respiration. In the modern hospital, where such functions can be taken over artificially, death means something more subtle: the absence of consciousness; the absence of activity in the brainstem, where automatic functions like breath and circulation begin; the absence of any hope of recovery. Machines can keep a heart beating and lungs pumping almost indefinitely, but a brain-dead person is a simulacrum of a living being, a body that persists despite the fact that whoever animated the body is already gone.

Many people found it difficult to accept that someone who was breathing, whose kidneys, perhaps, still worked and whose liver still functioned, even someone whose eyes opened and tracked your movements around the room, could nonetheless be dead.

In early 1968 Steptoe, then fifty-five, was in the audience at a London meeting of the gynecological section of the Royal Society of Medicine, hoping to get some tips on the use of laparoscopy in the treatment of polycystic ovarian syndrome. The speaker said laparoscopy simply did not work in visualizing the ovaries, that the only way to see them was with x-rays. Steptoe knew the speaker was wrong and suddenly, before he realized what he was doing, he was on his feet. "Nonsense!" he shouted in his deep, rumbly voice. The room of two hundred gynecologists went still. "Absolute nonsense!" he said. "You're quite wrong. I carry out laparoscopy regularly each day, many times a day." He asked permission to show slides of the polycystic ovaries he had visualized through his laparoscope.

After Steptoe's dramatic — and convincing — presentation, a tall, shaggy-haired man with piercing eyes approached him in the hallway outside. "You're Patrick Steptoe," the man said. "I'm Bob Edwards." And he began to describe his plans to treat women whose blocked fallopian tubes had rendered them infertile: give them hormones to stimulate ovulation, collect their mature eggs through laparoscopy, mix those eggs with their husbands' sperm in petri dishes, culture the fertilized eggs for a few days, and return them to the women's wombs through another laparoscope, thereby bypassing the blocked tubes.

"He was a scientist, I was a doctor," Steptoe later recalled. "We both wanted to help people who had seemingly insoluble infertility problems. So why not?"

Edwards and Steptoe knew it wasn't going to be easy. Even the first step — retrieving mature ova — was a feat that most surgeons could not manage. When a human egg matures naturally — usually just one per month — it forms a little pouch on the surface of the ovary, hanging there like a ripe fruit, squishy and retractable when a surgeon tries to grab it with a scalpel or a syringe. But Steptoe seemed to have a genie's touch. He had figured out how to use the laparoscope to take hold of the devilish little berry. He would insert the fiberoptic tube into his patient's abdomen to see where he was going, then insert a second laparoscope, through which he threaded a minuscule pair of scissors, with which he could snip off the mature egg follicle in the least disruptive way. It was an operation that left only a small or even an invisible scar (Steptoe usually tried to enter right through the belly button), and it was a technique that could be mastered, at least by a select few.

The next step, getting the sperm to fertilize the ovum in a petri dish, was elusive, too. Edwards and Steptoe worked at it for months, first by refining what Edwards called the "magic fluid" in which the eggs were cultured, then by varying the length of time between the harvesting of the egg and the introduction of the sperm. Still, fertilization seemed a long way away.

Also remote was the likelihood that the general public would approve of what they were doing. In 1969 a Harris poll found that more than half of American adults believed the emerging reproductive technologies — meaning in vitro fertilization, artificial insemination, and surrogate mothering — were "against God's will" and would "encourage promiscuity." More than two-thirds thought these techniques would signify "the end of babies born through love." Public sentiment was similar in Great Britain, where many people thought that fertilizing a human egg in the laboratory was crossing some critical line. As an editorial in the *Times* of London put it, "It is the implications rather than the direct consequences of the first successful fertilization of a human egg in a test-tube that pose the most searching moral problems."

And then, at last, success. On the afternoon when the British scientists looked under the microscope and saw a single sperm starting to penetrate the protective corona around the egg — when they finally witnessed the moment of fertilization — the babble of criticism that accompanied their every announcement momentarily stopped crackling in their ears. For Steptoe, a busy physician who rarely had time to stop and consider the mystery of life, the event was an almost religious experience. "I am one of the three or four people in the world who has genuinely seen this," he said, awed by the staccato rhythms of what Shettles liked to call the "dance of love."

Quietly, Landrum Shettles was forging ahead on work similar to Edwards and Steptoe's, in New York, in quarters just as cramped as theirs, his odds against success just as long. According to his own accounts, all issued several years after the fact, he successfully carried out a series of fertilizations, and, as he later claimed, that one implantation. Did he have a mystical feel for the egg that his British competitors lacked, an ability to intuit the environment it needed to grow and thrive? How else to explain why Shettles seemed to be doing so well, working alone and in secrecy, while a pair of British scientists with better funding and wider connections had failed over and over again?

"You certainly started the ball rolling with your blastocyst," wrote Shettles's friend John Rock of Harvard in 1970. "I wish it would not roll over some people so completely that they lose their wits entirely."

During this period, after the supposedly successful embryo transfer but before any public announcement of his success, Shettles was writing books. His first, which had been published in 1960, was the slim volume *Ovum Humanum*, a photographic atlas meant for scientists, physicians, and medical students, showing the human egg in all stages of development. Few embryologists of the day were as skilled as Shettles in creating photographs that so crisply revealed the gorgeous, incomparable unfolding of that early stage of human life. Shettles had a real flair for photography. He used a Leica camera with an attachment that allowed him to photograph images of the egg from a Zeiss phase contrast microscope, and he worked far into the night in the darkroom at Presbyterian Hospital. The images in *Ovum Humanum* had a rare clarity, looking less like biological specimens than like a swirling succession of birds' nests, mandalas, the Milky Way, and one or two elaborately crafted canapés. One photograph, of a blastocyst ninety-six hours after fertilization, looked uncannily like a man's face, crowned with cloudy white curls, wearing spectacles, and with an askew, sardonic pursing of the lips — not unlike Shettles's own expression in photos.

Over a period of six years, according to the text of *Ovum Humanum*, Shettles had photographed more than one thousand human eggs. Photographs from the book were reproduced in such textbooks as *The Book of Popular Science* (Grolier), *Animal Biology* (Harper & Row), *Human Genetics and Its Foundations* (Reinhold), and *Good Health* (W. B. Saunders), as well as on the cover of the British *Dictionary of Biology.* Even the archdiocese of New York, whose leaders fervently opposed the research in which Shettles was a pioneer, requested permission to reproduce one of his photos of a fertilized egg in the 1963 edition of the *Catholic Youth Encyclopedia.*

The New York cardinal might have liked Shettles's photos, but the Vatican remained staunchly opposed to his research. On July 25, 1968, Pope Paul VI issued the encyclical letter *Humanae Vitae* in which he reiterated the Church's belief that procreation must be connected with "the conjugal act," and, conversely, that the conjugal act must be connected with procreation. "The question of human procreation," wrote the pope, "like every other question which touches human life, involves more than the limited aspects specific to such disciplines as biology, psychology, demography or sociology. It is the whole man and the whole mission to which he is called that must be considered: both its natural, earthly aspects and its supernatural, eternal aspects."

While the pope's argument was explicitly aimed at birth control, it was an easy matter to apply it to the flip side of artificial contraception — artificial conception, in particular in vitro fertilization. "Unless we are willing that the responsibility of procreating life should be left to the arbitrary decision of men," he wrote, "we must accept that there are certain limits, beyond which it is wrong to go."

Almost in passing, he gave infertile couples a papal dispensation of sorts: his permission, not to use reproductive technology in their quest for children but, in this singular exception, to separate conjugal union from procreation — in other words, to continue to make love even though they knew no offspring would result. As he put it, "The sexual activity, in which husband and wife are intimately and chastely united with one another, through which human life is transmitted, . . . [does not] cease to be legitimate even when, for reasons independent of their will, it is foreseen to be infertile." Even if sex doesn't fulfill the procreative function for such couples, he said, it does fulfill its second purpose: as an expression of the love and unity of husband and wife.

In the early 1970s Shettles was a coauthor of two popular books about reproduction. The first, *From Conception to Birth: The Drama of Life's Beginning*, he wrote with Columbia radiologist Roberts Rugh and science writer Richard N. Einhorn. The photo of Rugh, whose eyepatch made him look a bit like Moshe Dayan of Israel, lent an air of rakish daring to the book jacket. Each new zygote, the authors wrote, "is a chance composite of two distinct lines of inheritance and is thus in every way unique, different from all other individuals. Since each child possesses a heritage from the past and is a link with the future, this privilege of life is to be cherished, guarded, and nourished with all the modern resources and knowledge available. In such manner life on earth can be improved." The book featured elegant line drawings of prenatal life as well as otherworldly photographs, most taken by Shettles or Rugh, of fetuses floating facelessly in a strange and silent solitude. Even in direct competition with Lennart Nilsson's more famous photographs from inside the uterus, *A Child Is Born*, Rugh and Shettles's book went into more than a dozen printings.

Even more successful was *Your Baby's Sex: Now You Can Choose*, in which Shettles and his coauthor, David M. Rorvik, explained how a couple could tip the odds of conceiving a child of the desired sex.

With the Shettles technique, they wrote, 75 percent of couples trying for a girl and 80 percent of those trying for a boy could conceive a child of the sex they wanted. They advised readers that if they wanted a girl, for instance, the woman should douche first with a vinegar solution, should be flat on her back during intercourse, and should refrain from having an orgasm.

The book was a hit. *Your Baby's Sex* remained in print for more than thirty years and eventually sold more than a million copies, even though it touted a method that some said was no more reliable than a coin toss. And for a while it made Shettles one of the most famous obstetrician-gynecologists in the country — as well as one of the most audacious.

David Rorvik, Shettles's coauthor, was a freelance science writer who was able to use his contacts as a journalist to publicize the book in some of the leading magazines of the time. On April 21, 1970, for instance, *Look* put a photo of a chubby, naked baby on the cover with the heading "How to Choose Your Baby's Sex — A distinguished doctor offers parents a safe, simple method that works." Rorvik himself wrote several magazine articles (including three in *Family Circle* alone) about the Shettles method, sometimes failing to mention that he was the coauthor of a book promoting that very thing. In a long article for *New York* he adopted the magazine's studied hipness, calling Shettles a "gynodynamo." The article, accompanied by a rare photograph of Landrum Shettles smiling, made the doctor look as if he was just too cool to be bothered with issues such as ethics, issues that concerned some of his slower, more plodding colleagues. Did Shettles have any qualms about creating test tube babies? Rorvik asked in the article. No, said Shettles: "I have none. I always told my critics I'd like nothing better than to grow a beautiful lab assistant." That kind of smart-ass comment wasn't typical of Shettles, who usually peppered his speech with Mississippi farm-boy exclamations like "Lordy" and "Land sakes." But it helped turn Shettles into a celebrity — and into an annoyance for his department chairman.

Shettles's fame crested in May 1971, when *Look* magazine ran a

3
LAUGHINGSTOCK

> When you have eliminated the impossible, whatever remains, however improbable, must be the truth.
>
> — SIR ARTHUR CONAN DOYLE, "The Sign of the Four" (1890)

JOHN DEL-ZIO DIDN'T THINK it odd, really, to be meeting the illustrious Landrum Shettles in a hallway at Presbyterian Hospital rather than in his office. There was something so peremptorily awkward about Shettles's social style that no single aspect of it stood out; the gestalt was odd enough. But in fact, the reason Shettles did not meet Del-Zio in his office was that he didn't truly have an office.

Despite his long association with Columbia and with Presbyterian Hospital, by 1973 Shettles had been stripped of most of his duties in patient care. To hear his chairman, Raymond Vande Wiele, tell it, Shettles wouldn't even have had a job there at all were it not for the kind patronage of Vande Wiele himself.

Vande Wiele, then fifty and just one year into his tenure as Columbia's chairman of obstetrics and gynecology, was supposed to get rid of Landrum Shettles, whom the hospital's administrators saw as a drain on resources as well as a public embarrassment. The kind of attention he generated — writeups in general-interest magazines like *Look* and *Esquire;* articles in tabloid newspapers about his sex-selection method; his pale, delicate-featured face, usually unsmiling, appearing in publications across the country — was not the kind of attention that a staid place like Columbia considered appropriate. Most university administrators have a complex attitude toward publicity: while on the one hand they love seeing the institution's name in the

cover story called "The Test Tube Baby Is Coming," also written b
Rorvik. On the cover, beneath the thirty-five-cent newsstand pric
and the familiar block-letter logo, was a photograph of what seeme
to be an embryo in aspic in the palm of a flat, broad surgeon's hand
The embryo, seven weeks old, had been removed during a hyster
otomy and kept alive by immersing the entire amniotic sac in a bath
of oxygenated saline solution to keep the heart beating for severa
hours. That competent surgeon's hand belonged to the man who had
devised the way to keep the heart beating, the man who had taken the
photograph with his right hand while holding the embryo in his left
— Landrum Shettles.

In the *Look* article, Rorvik described Shettles as nothing less than the Galileo of the twentieth century. He wrote that Shettles had singlehandedly elevated the clumsy science of in vitro fertilization "to an art." He called the doctor a visionary, a pioneer, an unappreciated genius, "the first man in the world to witness the drama of human fertilization." He made Landrum Shettles famous.

Indeed, he was so famous that some people thought Shettles's colleagues, especially his boss, were getting just a little bit jealous.

newspapers, which they hope will bolster its reputation and, especially, increase donations, on the other hand they regard publicity as just a bit unseemly for the ivory tower. Yes, it was nice to have Columbia's name associated with one of the most famous baby makers in the city — but did he have to be quite so . . . well, quite so famous?

So when Vande Wiele became department chairman, the administrators told him to fire Shettles. Uncharacteristically, Vande Wiele refused. He did not have the heart to fire a once-promising colleague who was so close to retirement and who was responsible for the private school tuition and upkeep of seven children. Vande Wiele himself was the breadwinner of a family of three girls, and he had great respect for the role of *paterfamilias* — even if Shettles's version of that role was unlike just about anybody else's. Shettles lived alone, apart from his wife and children, in a hospital cubbyhole. Tropical fish tanks along one wall emitted a soothing hum and an eerie, bluish glow; above the fish was a collection of clocks, each ticking out its own idiosyncratic time. Papers and journals were stacked everywhere, and standing upright in one corner was an old cot that Shettles would pull down when it was time to sleep. He chose this claustrophobic way of life over the more typical domestic version, being crammed into an Upper West Side apartment with his huge family. No one, least of all his seven children, ever figured out why Shettles chose to live the way he did, except perhaps to keep himself immersed in and cozily enveloped by his work.

Vande Wiele devised a way to demote Shettles rather than fire him, hoping to mollify his superiors and at the same time assuage his own guilt. He put Shettles in charge of admissions triage at the obstetrics and gynecology clinic, a job usually given to a nurse, a task Vande Wiele thought he could manage without causing too much disruption.

Despite this gesture of solidarity, Vande Wiele was far from a Shettles partisan. Whenever Shettles made a pronouncement in a scientific or lay publication, Vande Wiele would write to ask him to back it up with data; whenever Shettles was quoted in a magazine interview, Vande Wiele would write to ask him where the reporter could have gotten the idea that Shettles would say such a thing. Again and again he wrote letters to Shettles, carefully keeping the carbon copies in his files. He repeatedly asked for some experimental evidence for Shettles's claims regarding sex selection or laboratory fertilization, pleaded with him to stop making those claims, and especially begged him to stop talking to reporters.

Raymond Vande Wiele was a proper man intent on doing things properly. He could be a creative scientific mentor — "think again" was his favorite phrase, used to prod colleagues to envision daring routes out of research dead ends — but as an administrator he was cautious, sometimes overly so. He spent much of his time toeing the line and making sure his faculty did as well. This was how he approached experimentation on humans, an aspect of biomedical research in which toeing the line was especially difficult because the line was so hard to discern.

In 1973 there weren't many signposts indicating how research with human subjects should be done. The primary guide was the Declaration of Helsinki, offered by the World Medical Association in 1964. Privacy and integrity were the main focus of the document, which stated that scientists should take "every precaution" to "respect the privacy of the subject and to minimize the impact of the study on the subject's physical and mental integrity and on the personality of the subject." The Helsinki Declaration had no means of enforcement, but it outlined ways in which some other regulatory body could ensure that research on human subjects was conducted in an ethical manner: prior animal research, a competent investigator, oversight by a local review committee, and "freely given informed consent." Its overriding ethical principle was that "biomedical research involving human subjects cannot legitimately be carried out unless the importance of the objective is in proportion to the inherent risk to the subject."

A document with more regulatory teeth came in the form of a policy and procedure order issued by the U.S. Department of Health, Education, and Welfare (HEW) in 1966. This order formalized a code of behavior for clinical research, defined as studies that involved "any medical or surgical procedure, any withdrawal or removal of body tissue or fluids, any administration of a chemical substance, any deviation from normal diet or daily regimen, [and] any manipulation or observation of bodily processes, behavior or environment." The HEW policy, which included many of the same principles as the Helsinki Declaration, was enforceable only for research conducted under federal grants. It had no authority over privately funded experiments.

Another mechanism of oversight at the time was the "letter of assurance" that a university had to sign before it could receive any form of federal research support. Columbia University's College of Physicians and Surgeons had on file one such letter, which stated that all scientists at the medical school promised to abide by HEW guidelines for human subject research, whether or not their particular study was being conducted under a federal grant. Columbia promised to implement a three-step procedure for reviewing faculty members' plans to do research on human subjects: such a study had to be approved first by the department chairman, then by the department human experimentation committee, and finally by a larger human experimentation committee, which reviewed research proposals from the entire medical school. Any violation of this procedure, in Vande Wiele's view, meant that the medical school would run the risk of losing every cent of its government funding — a loss that would no doubt be measured in the tens of millions of dollars.

With so much at stake, Vande Wiele was careful about what kind of clinical research he would and would not allow, especially when that research meant using tissue from human eggs, embryos, or fetuses, since abortion opponents had turned such manipulations into political land mines. That is partly why he was so rankled by Shettles's media pronouncements about his studies in IVF and sex selection.

Vande Wiele also found it personally embarrassing to be associated with Shettles, and he did what he could to distance himself, and the entire department, from Shettles's public persona. "It is annoying enough, wherever I go, to have to listen to the pious condolences of the others about the sad state of affairs of our department," he wrote to a colleague at Columbia. "It would be even worse if we would let Dr. Shettles make us the laughingstock of the academic community."

At one point Vande Wiele had taken the extreme step of sending out a formal statement to the press stating that Shettles's method of sex selection had no connection with Columbia. The statement read:

> A number of claims have been made in the lay press related to the possibility of influencing the sex of a baby, by various preconceptual measures to be taken by the prospective parents. In most of these articles reference was made to the work of Dr. Landrum B. Shettles of the College of Physicians and Surgeons, Columbia University. Dr. Vande Wiele, Chairman of the Department of Obstetrics and Gynecology at the College of Physicians and Surgeons, when contacted, indicated that he was unaware of any work in his department by either Dr. Shettles or any other members, to back up these new studies. Dr. Vande Wiele expressed his extreme concern about these references to the Department of Obstetrics and Gynecology, since he feels that reliance on the methods suggested in these articles could only lead to bitter disappointments.

In May 1971 Vande Wiele had been particularly angered by the *Look* article featuring Shettles's hand on the cover. In the article, Rorvik placed Shettles far out front in the international competition to perform human IVF, implying that he would be making a test tube baby any day now. At the time it did indeed look as if Shettles might have pulled ahead in the race, because his leading competitors in England had suffered a serious setback.

The force that so often drives scientific research — government grants — seemed to be drying up for Edwards and Steptoe. They had applied to the Medical Research Council (MRC) of Great Britain for funding, and in 1971 the council turned them down with sobering finality. "I am sorry to have to tell you," the council secretary wrote to Edwards, "that . . . the proposed investigations in humans . . . were considered premature in view of the lack of preliminary studies on primates and the present deficiency of detailed knowledge of the possible hazards involved." Nonetheless, he added, if Edwards was interested in conducting the research on monkeys, the council "would be prepared to consider an application."

Should government money dictate the direction of research in this way? Whether or not it should, it most certainly does. The "war on cancer," for instance, which President Richard Nixon declared in 1971, was fought almost entirely by directing more and more appropriations to the National Cancer Institute (NCI), which became the fastest-growing institute at NIH. Between 1970 and 1980 the NCI budget grew four and a half times while the overall NIH budget only tripled. Because applications directed to the NCI seemed to stand a better chance of getting approved, a significant number of investigators tried to make the case, in their grant applications, that their research could one day lead to a cure for cancer. Despite these heavily funded battles, however, victory still eludes us more than three decades after the war on cancer was declared.

But just as the government can encourage a certain line of research by throwing extra money its way, so can it choke off research by refusing to fund it altogether. That seemed to be what Britain's Medical Research Council was trying to do with IVF.

Besides being unhappy about losing the money, which he and Steptoe needed badly, Edwards was furious that the MRC was making the same argument he was hearing from other quarters: that he needed to do more animal research before moving to humans. He adamantly disagreed with this point of view; what could more mice and monkeys teach him about in vitro fertilization that he didn't already know? To Edwards, lab animals were poor models for human reproduction. There were similarities, of course, between the reproductive system of the mouse — Edwards's original specialty, and one of the most commonly studied animals in embryology — and that of the human. The mouse, like the human, is a mammal — which means, among other things, that the eggs are fertilized inside the body, the fetuses grow inside a uterus, and the young are fed after birth by the mother's milk. And, as with humans, mouse fertilization occurs in one of the two oviducts, the tubes connecting the ovaries to the uterus, with the fertilized egg taking several days to travel down the oviduct to the womb. (In humans the oviducts are also known as the fallopian tubes, named after the seventeenth-century Italian anatomist Gabriele Fallopio, who discovered them.) As the fertilized egg travels down the oviduct, it undergoes cleavage, splitting into new cells without getting any bigger overall. At the blastocyst stage, in mice as in humans, the developing embryo has completed its trip down the oviduct, reached the uterus, and implanted itself in the uterine wall.

But in many ways the mouse can teach us nothing about how human fertility works — and it was the differences between mouse and human that most bedeviled Edwards. It takes just twenty days, compared to a human's two hundred and eighty, to grow a new mouse. Mama Mouse is receptive to the mating behavior of Papa Mouse during a single day of her five-day menstrual cycle. She is interested in sex only on the day she ovulates, while human females are sexually responsive throughout their cycles. And ovulation is different in the two species, too. In mice, it occurs in both ovaries at the same time, and a dozen or more eggs may come to maturity at once. In humans, eggs ripen in alternate ovaries from one menstrual cycle to the next, and usually only one egg is released at a time, with the occasional double ovulation leading to the possibility of fraternal twins. In mice, which are nocturnal, ovulation always happens in the evening; in humans it can take place at any time.

As for research on in vitro fertilization in other mammals, each species had its own reproductive idiosyncrasies. To fertilize rabbit eggs in vitro, the sperm had to be artificially capacitated. Hamster embryos that were frozen after two and a half days of cleavage grew into grossly deformed embryos, a change that was not seen in hamsters frozen at different points in development — nor was it seen in the embryos of any other species, no matter when they were frozen. And primates, despite their close relationship to humans, presented perhaps the greatest differences of all; in primates the cervical canal is so convoluted that transplanting a lab-fertilized embryo into a macaque or a chimpanzee would be significantly more difficult than it is in a woman.

Edwards had tried to explain these distinctions between animal and human reproductive systems at every opportunity. He had spent years insisting that it wasn't possible to know anything about how IVF would work in humans without just doing it in humans. Now the rejection from the Medical Research Council made it clear that no one had been listening. And it was clear, too, that nobody was going to make this easy for him. In the race to make a test tube baby, it seemed, there was no real alternative to slow and steady.

"Slow and steady" was never Shettles's approach. Nor was he careful and deliberative, according to Rorvik, whose *Look* article stated that Shettles never wondered about the viability or normality of a baby that might result from his experimentation; he didn't really consider such matters to be any of his concern.

It was this last bit of arrogance that most angered Raymond Vande Wiele. "I think you must admit that I have some reason to be fed up with all this business," he wrote to Shettles when the *Look* article appeared. "If Mr. Rorvik is trying to kill you professionally, he couldn't do a better job."

4
Out of Control

What are we going to do with the mistakes?

— James Watson, "Fabricated Babies" (1971)

The Kennedys of Massachusetts could focus media energy on just about any issue that captivated them. That is what happened when the Kennedy Center for the Performing Arts opened in Washington, D.C., in September 1971 — and what happened, too, at a conference held in that massive space just one month later, on the previously unsexy subject of biomedical ethics.

As the opening night ceremonies approached, newspapers and magazines were filled with analyses of dance, theater, and symphonic music — not typical fare in those days for the *Washington Post* and *Washington Star.* The Kennedy Center's unveiling on September 8 brought the city as close to being chic as it had been since John F. Kennedy was president. Though it was then the dour days of the Nixon administration, the attendees sparkled with a glamour that was purely Kennedyesque.

The event was a cultural landmark, fostering the belief that the performing arts center's very existence would transform Washington from a company town that folks abandoned on the weekends into a true destination, a city of intellectual *gravitas.* It featured the premiere of Leonard Bernstein's *Mass,* which required a large pit orchestra, two full choruses, a boys' choir, a Broadway-sized cast, a ballet company, a rock group, and a marching band. *Mass* proved to be a magnet for Middle America's anger, which was a complete surprise to Bernstein, shielded in his New York intellectual cocoon, far from the delicate nerve endings of religious fundamentalists from the nation's heartland. He had never anticipated how deeply offended some people would be by this powerful, eclectic, but often dissonant musical offering, built entirely around the Christian liturgy, composed by a blithely atheistic Jew. President Nixon himself was suspicious of Bernstein's libretto. Several members of the White House staff, whom Nixon had assigned to listen in on the *Mass* rehearsals, reported back that they believed "coded messages" were embedded in the Latin text.

That opening night evoked the Camelot of the Kennedy years, one brief, shining moment when Americans believed — somewhat naively — that mankind was perfectible, that knowledge was power, and that society could use its myriad tools to make the world a better place. But it also evoked, for those who were paying attention, something darker: a deepening divide in American culture, with conflicting religious and moral sensibilities carrying people in directions entirely unanticipated, and unrecognized, by those involved in the secular pursuits of the arts and sciences.

One month later the new performing arts center was the site of a most unusual film premiere — and once again the Kennedy family was in charge. The film *Who Should Survive?* was a documentary produced at Johns Hopkins Hospital about a baby born there with Down's syndrome. The newborn also had a severe intestinal blockage. It could be surgically repaired, but, if left untreated, it would kill him within weeks. His parents decided not to operate. They said they preferred to allow their baby to die, believing that it would be unfair to their other two children to bring home a retarded child.

Who Should Survive? charted the hospital staff's anguished withholding not only of surgery but also of food and water. As instructed by the parents, the doctors and nurses placed the baby in a quiet corner of the nursery and watched, seeing but trying desperately not to see, as he slowly starved to death. It took fifteen days.

The screening, on October 16, was part of an all-day conference sponsored by the Joseph P. Kennedy, Jr., Foundation, which was devoted to analyzing contemporary issues in bioethics. Named in memory of JFK's older brother, who died in battle in World War II, the foundation was directed by Eunice Shriver, the third oldest Kennedy sister. Almost all of the surviving members of that generation were on the board of directors. The foundation focused on the rights of the mentally retarded, spurred by the fact that the oldest surviving Kennedy sibling, Rosemary, then fifty-two, was retarded. But the foundation took as its responsibility a whole range of bioethical issues, including the promotion of the field of bioethics itself, which was just starting to achieve some academic cachet. That same month the foundation had provided $1.35 million in start-up money for the nation's first university-based institute for bioethics, the Joseph and Rose Kennedy Institute for the Study of Human Reproduction and Bioethics at Georgetown University, just up the Potomac River from the Kennedy Center.

Bioethics at that time was part philosophy, part theology, with some biology, medicine, and law tossed into the mix. The Kennedy Institute was developing a graduate program, in collaboration with the Georgetown philosophy department, that would produce Ph.D.s with special expertise in bioethics. But it was a lonely enterprise; the only other American institution focusing on bioethics at the time was the Institute of Society, Ethics, and the Life Sciences, nicknamed the Hastings Center for the town where it was located, fifteen miles up the Hudson River from New York City. Under the direction of cofounders Willard Gaylin and Daniel Callahan, the Hastings Center was trying "to fill the need for sustained, professional investigation of the ethical impact of [the] biological revolution." It published a bimonthly report, worked with universities and medical schools to have ethics courses added to their curricula, and was just starting to make its mark on the conference circuit.

The unstated theme of the Kennedy Foundation conference was that scientific progress, for all its promise, could also have a dark side, and that it behooved us to educate ourselves so we could keep the dark side from eclipsing the good. It echoed the theme of the year's surprise bestseller *Jonathan Livingston Seagull*, which took its place alongside the year's most successful books, including *I'm OK, You're OK*, the granddaddy of the self-actualization movement, and a book about a new high-protein, low-carbohydrate diet, *Dr. Atkins' Diet Revolution*. In the novel, Jonathan is a seagull who spends all his time trying to find out how the world works, how he works, what flying is all about. In other words, he thinks like a scientist. "For a thousand years we have scrabbled after fish heads," Jonathan tells his fellow seagulls, "but now we have a reason to live — to learn, to discover, to be free!" For his attempts to learn and to discover and for his refusal to accept that "life is the unknown and the unknowable," the elders of the flock cast him out. Trying to know the unknown, they tell him, is "reckless irresponsibility."

In his opening talk at the Kennedy Foundation conference, Senator Edward M. Kennedy of Massachusetts, the foundation's president, also warned of the potential recklessness of trying to know the unknown. Humanity's unprecedented ability to determine its own destiny raised significant bioethical questions, he said. "Without answers to these questions, man is, in the deepest sense, out of control."

People were just starting to talk — and fret — about creating life in the lab, as evidenced by Rorvik's article in *Look* with Shettles's hand on the magazine cover. That same month two articles about the implications of IVF appeared in the more highbrow *Atlantic Monthly*. But such articles were still rare; few people were discussing the ethics of assisted reproduction in 1971, either in print or around the dinner table. Most Americans did not even know what forms of assisted reproduction were possible; "test tube babies" still meant babies born after artificial insemination, and even that procedure was obscure to the vast majority. Around the time of the Kennedy Foundation conference, when some twenty thousand babies a year were being born in the United States as a result of artificial insemination, a Harris poll revealed that only 3 percent of Americans even knew what it was.

If reproductive technology was a concern, then, it was still a rather rarefied concern, something discussed over cigars and brandy rather than pretzels and beer. Robert Edwards traveled in these elite circles; Landrum Shettles did not. Edwards's British accent sounded to American ears impeccably upper class (even though it reflected his modest Yorkshire roots), while Shettles's high-pitched stuttery speech was dirt-farm Dixie. Edwards befriended leading IVF researchers around the world, while Shettles was close to no one, so isolated in his work that he kept no lab notebooks, the common currency of scientific collaboration, insisting that he had all the details safely inside his own head. Edwards's work was described, in sometimes melodramatic tones, in those cover stories in the *Atlantic;* Shettles's name was mentioned only in passing. And Edwards was invited to the Kennedy Foundation conference in October, while Shettles stayed in New York, fussing with his petri dishes.

The conference felt like history in the making, perhaps because the Kennedy Center's red carpets and crystal chandeliers implied momentousness, even at nine o'clock on a Saturday morning. Maybe it was all those Kennedys, or the fact that 1,200 people had signed up to attend. Maybe it was the list of panelists and moderators, which included some of the most famous names of the time. It was a collection not just of Nobel laureates (Jacques Monod, Joshua Lederberg, Sir John Eccles) and foundation and university presidents (Philip Handler, John Knowles), but also of cultural stars and literary figures, including William Styron, Germaine Greer, B. F. Skinner, and Norman Podhoretz. Mother Teresa, the Calcutta relief worker, was there, as was author, activist, and Holocaust survivor Elie Wiesel. And each panel was moderated by a famous television news anchorman or commentator, including John Chancellor, Frank McGee, and Roger Mudd.

"The crowd was modish," said a radio reporter covering the meeting. "It included a healthy proportion of the exuberant young that make Washington an increasingly interesting place to live. And the kind of people that always appear at any event linked to the Kennedys were there. There were students from nearby colleges and universities, a few harried journalists, and the participants themselves — a dizzying mélange."

After the screening and discussion of the twenty-six-minute film *Who Should Survive?* hundreds of the conferees moved out into the tart blue October day. They shuttled on buses to the Shoreham Hotel, where they signed up to listen to experts on various panels discussing "Who Should Be Born?" "The Modification of Human Behavior," and "Why Should People Care?"

The session on in vitro fertilization was called "Fabricated Babies." It featured two of the field's leading scientific figures and two of its staunchest opponents. Also on the panel was one of the most famous scientists in America, James Watson. The first IVF expert was Edwards, making a quick stopover on his way to deliver a lecture in Japan. The second was Howard Jones, who believed he had been invited only because Sargent Shriver (Eunice Shriver's husband), with whom he had a passing acquaintance, had enlisted him to persuade Edwards to come. Having asked that favor, how could Shriver not extend an invitation to Jones as well?

Edwards had been hesitant about attending the Washington conference. His life was one of constant travel: Steptoe was having such success harvesting eggs that Edwards was making the five hundred-mile round trip from Cambridge to Oldham almost every week to do fertilizations, and he was speaking at meetings all over the world to announce his slow and steady progress and to see what other people were up to. He did not really want to add to the strain of the grueling trip to Japan, but when Jones called to ask him to attend the meeting, Edwards agreed. Although they had barely seen each other since their collaboration six years earlier, that single intense summer, with its successes, setbacks, and discoveries, had solidified a friendship between the two IVF pioneers that would sustain them through a succession of challenges — beginning that very afternoon.

One of the IVF opponents on the "Fabricated Babies" panel was Leon Kass, who had just left a research job at the National Institutes of Health. Although trained as a biologist, Kass's dominant interest was ethics, an interest sparked by a *Washington Post* article by a fellow scientist who had written about the prospect of human cloning in a manner that Kass found disturbingly flip. Kass's first formal position as an ethicist was at the National Academy of Sciences, a professional society that accepts assignments from Congress and the executive branch to consider complex scientific developments and their implications for society. At the time of the Kennedy Foundation conference, Kass was executive secretary of the Committee on the Life Sciences and Social Policy, which was working on a report on "technology assessment" of four emerging technologies, including IVF.

A prolific and forceful writer who was more didactic in print than he was in person, Kass had already gone on record denouncing in vitro fertilization. He called it part of a "new holy war against human nature," which would lead to "the divorce of the generation of new human life from human sexuality and ultimately from the confines of the human body, a separation which began with artificial insemination and which will finish with ectogenesis, the full laboratory growth of a baby from sperm to term." To Kass the slope was not only slippery, it was downright friction-free, moving from laboratory fertilization to complete laboratory gestation with a grim inevitability. He saw IVF as the first step on that slope, and therefore a threat to everything he held dear: the sanctity of the nuclear family, monogamy, fidelity, all the ties that bind families and generations.

IVF also threatened our equanimity, Kass wrote, by moving "the mysterious and intimate processes of generation" out of the darkness of the womb and into the bright light of the laboratory, "beyond the shadow of a single doubt." Such harsh, cruel knowledge could be destabilizing, more information than we would know what to do with. "The first and perhaps best attack on this process of enlightenment was delivered by Sophocles," he wrote, referring to the ancient Greek playwright's treatment of the myth of Oedipus, who tried to avoid his destiny of killing his father and marrying his mother, only to discover to his horror that he had in fact lived it. "Do we mean to ignore what we learned from Oedipus' prideful efforts to bring into full and public light the mystery of his origins?"

The other prominent IVF critic on the panel was Paul Ramsey, a professor of religion at Princeton. Ramsey believed IVF research was inherently unethical for one irrefutable reason: the individual most likely to be damaged, the embryo, could not possibly give informed consent. "We cannot rightfully get to know how to do this," he wrote, "without conducting unethical experiments upon the unborn who must be the 'mishaps' (the dead and the retarded ones) through whom we learn how."

So there they were: Edwards and Jones on the pro-IVF side of the aisle, Kass and Ramsey opposite. The meeting organizers couldn't have set things up for a wrangle any better had they invited Galileo to debate the pope.

Edwards spoke first. As was the tradition at meetings like this, he gave a summary of the work he had done up to that point. He repeated what he had announced in *Nature* and *Scientific American:* that his partner, Patrick Steptoe, had perfected the tricky noninvasive surgical technique of laparoscopy and could now use it reliably to retrieve mature eggs from his patients; that Edwards and Steptoe had treated forty-nine women with fertility drugs to stimulate their egg production, retrieving a total of twenty-nine human eggs through laparoscopy; that they had placed those eggs in petri dishes and incubated the cultures for one to four hours. When the eggs were sufficiently mature, the scientists had introduced washed samples of spermatozoa into the petri dishes, hoping to achieve fertilization; they had been successful in twenty-three cases, and the resulting pre-embryos had lived for up to two days in the laboratory. But as of the date of the conference, he reported, none of the twenty-three externally fertilized zygotes had yet developed into blastocysts ripe for implantation in the egg donor's uterus. And no such implants had yet been tried.

Edwards delivered his summary in typical hurried style, evidence of a mind in high gear. His words came in an insistent rush, his voice so whispery and thickly accented that the audience had to strain to hear him. Perhaps it was Edwards's steady exhalation of progress reports that set Ramsey off, making him more stridently anti-IVF than usual. Or perhaps Ramsey had intended all along to take the offensive that afternoon.

"In vitro fertilization constitutes unethical medical experimentation on possible future human beings, and therefore it is subject to absolute moral prohibition," Ramsey proclaimed, using an oratorical overdrive that led one observer to use terms like "prickly," "pontifical," and "formidably disputatious" when talking about Ramsey's argumentative style. He buried IVF with the slippery-slope argument — or, in his phrase, "the thin edge of the wedge." In his view IVF paved the way for obscene genetic manipulations straight out of the novel *Brave New World.* Using Aldous Huxley's own terminology, he said it brought us a "long step" closer to the use of "hatcheries," where babies were grown entirely in laboratory dishes, as well as to "the introduction of unlimited genetic changes into human germinal material while it is being cultured by the Conditioners and Predestinators of the future."

Peering down from the podium, Ramsey mocked Edwards's contention that IVF was a humane medical treatment for desperate infertile couples. How could it be a "treatment"? After all, it cured nothing. "Is the 'clinical defect' of infertility remedied by in vitro fertilization?" he asked. "I should say not! Instead, the child as a product of technology is to be brought forth, without remedying the woman's infertility. She remains as infertile as before."

To Ramsey's way of thinking, the only true treatment for infertility was reconstructive surgery of the fallopian tubes. When the surgery worked, he said, *that* was when an infertile woman was no longer infertile. If you merely gave her a baby using IVF, then she was just an infertile woman with a baby. "Without curing that condition, in vitro fertilization concentrates on a product," he said; "it is therefore manufacture by biological technology, not medicine."

In a startling conclusion, Ramsey said he hoped that the first test tube baby would be born so badly damaged that the whole enterprise would be brought to a halt. "I do not actually believe that the good to come from public revulsion in such an event would justify the impairment of that child," he hastened to add. "But then for the same reason, neither is the manipulation of embryos a procedure that can possibly be morally justified — even if the result happens to be a Mahalia Jackson."

Ramsey's holy-rolling wrath and his use of the staid academic lectern as a pulpit were unusual in a meeting like this. Kass's manner was more typical. His concern was just as grave, but his style was that of a moderate, thoughtful scholar, not an evangelical preacher. Kass began by saying that while he believed the social effects of IVF were pernicious enough — withering or even severing the traditional ties that bound nuclear families — he was just as worried about the procedure's likely medical effects. Test tube babies would probably be damaged babies, said Kass, or at least had to be presumed so. "It doesn't matter how many times the baby is tested while in the mother's womb"; scientists like Edwards "will never be certain the baby won't be born without defect."

Kass said he was reminded of the old joke about the airline pilot who made an announcement to his passengers: "Good afternoon, ladies and gentlemen. We are flying at an altitude of thirty thousand feet and at a speed of six hundred miles per hour. I have some bad news and some good news to report. The bad news is that we are lost. The good news is that we are making very good time." Not so different, according to Kass, from the current predicament of hurtling forward, full speed ahead, with IVF research, without a clue about where the research would lead.

Edwards sat quietly. Beneath a calm exterior, he was mentally dismissing his opponents' commentary as scarcely credible. So much of medicine involves steps short of cures, he noted, including such stopgap measures as insulin, spectacles, and false teeth. He concluded that Ramsey's tirade, despite the volume and the passion, was altogether specious. As for Kass's complaint that one couldn't be absolutely certain that IVF was safe — well, nothing in life was absolutely certain. And nothing about its safety would be known, no matter how many monkeys and mice were tested beforehand, until IVF was conducted on human beings. In addition, Edwards had heard the "thin edge of the wedge" argument before. "The whole edifice of their argument is fragile," he said later: "that nuclear physics led inevitably to the atomic bomb, electricity to the electric chair, civil engineering to the gas chambers. Surely acceptance of the beginning does not necessitate embracing undesirable ends."

Next up was James Watson, not really taking either side. In 1953, when Watson and his British colleague Francis Crick, working at Cambridge University, discovered the structure of DNA, they were acclaimed all over the world. When they shared the Nobel Prize for their accomplishment in 1962, their faces beamed toothily from front pages everywhere; they were fawned over and publicly adored. And when Watson published his rollicking, gossipy memoir, *The Double Helix*, in 1968, he became a true celebrity in a way that transcended his scientific reputation. By the time of the Kennedy Foundation meeting, Watson was notorious as a free thinker and as something of a loose cannon. He had testified before Congress earlier that year on genetic manipulation and the possibility of human cloning, and in May he had published an article based on his testimony in the *Atlantic Monthly*, an issue that also included a lengthy essay on "the obsolescent mother." Watson's article, because of both the stature of its author and the bluntness of its approach, had already become a cornerstone in the debate over reproductive technology among the East Coast intelligentsia — of which Watson, as a Nobel laureate and a tenured professor at Harvard, was a well-established member.

In the *Atlantic* article, "Moving Toward Clonal Man," Watson warned that scientists were spelunking into some crevices of nature that nonscientists might prefer to leave alone. But the public had no chance to weigh in with their opinions, he wrote, because the exploration itself was barely discussed, even among the scientists doing the exploring. The research simply went on unimpeded.

"Does this effective silence imply a conspiracy to keep the general public unaware of a potential threat to their basic ways of life?" wrote Watson. "Could it be motivated by fear that the general reaction will be a further damning of all science, thereby decreasing even more the limited money available for pure research? Or does it merely tell us that most scientists do live such an ivory-tower existence that they are capable of thinking rationally only about pure science, dismissing more practical matters as subjects for the lawyers, students, clergy, and politicians to face up to?"

Contrary to popular opinion, Watson wrote, humanity was not necessarily well served by the march of discovery: "Every scientific advance [does not] automatically make our lives more 'meaningful.'" And science did not necessarily progress in a linear, predictable way: "The belief that surrogate mothers and clonal babies are inevitable because science always moves forward . . . [is] nonsense."

Nevertheless, he said, the slippery slope was about to be traversed, with or without anyone's consent. Once the first test tube baby was born — sometime between 1980 and 1990 was his best guess — then the next step, human cloning, would probably be accomplished within the next fifty years — that is, by the year 2020. Would this be a good thing or a bad thing? Watson did not say. But he did have some fun wondering who should be cloned: Ringo Starr? Frank Sinatra? Raquel Welch? the shah of Iran? The list itself seemed to vindicate C. S. Lewis's warning that if we enshrine a certain set of ideals in a "preordained" gene pool, the resulting humans might seem, even a single generation later, ridiculously out of date.

In the article Watson did not argue either for or against IVF and cloning; he just wanted these matters discussed in the open, in public, so that the implications could be envisioned, planned for, and, if need be, short-circuited before the science took on a life of its own.

Now, when he loped to the podium at the Kennedy Foundation conference, all crazy hair and popping eyes, he was feeling puckish, more accusatory and inflammatory than he had been when he delivered his congressional testimony or wrote his *Atlantic* article. He turned to face Robert Edwards, looked him straight in the eye, and called him a baby killer.

"You can only go ahead with your work if you accept the necessity of infanticide," Watson told him. "What are we going to do with the mistakes?"

This warning, coming as it did from a widely admired scientist who could be expected to push for the free, unfettered pursuit of knowledge no matter where it might lead, was especially jarring. It was almost as though Jesus had spoken, telling his disciples not to be quite so devout.

After the other three speakers had finished, Robert Edwards moved to the microphone, relishing his chance to respond. The only sound in the room was the steady thrum of the air conditioner. "I accuse Paul Ramsey," Edwards said, "of taking up an ethical stance that is about one hundred years out of date." His voice now rose above its habitual whisper. "One that is totally inapplicable to meet the difficult choices raised by modern scientific and technological advance. Dogma that has entered biology either from Communist or from Christian sources has done nothing but harm . . ." The rest of his rejoinder was drowned out by a roomful of people on their feet, clapping. Amazed at the response, Edwards glanced down at Ramsey to see how this spontaneous outburst affected him. "I felt sorry for him," he later recalled. "His point of view had been shatteringly rejected by the audience. Indeed, he made no further contribution to that symposium."

Edwards made short work of the arguments offered by Kass and Watson as well. "We shall do our transplants and go on with our work as we decide, not as anyone else decides," he said flatly. The only panelist who spoke up in Edwards's defense was Howard Jones, who likened his friend's treatment by Kass and Ramsey to the trials of Galileo, who was arrested for his teachings about a heliocentric universe, which ran counter to Church dogma. Jones begged the audience not to let such an attitude seep into the twentieth century. "It's unthinkable," he said, "that there should be any area of scientific investigation set off limits."

Running off to catch his plane to Tokyo, Edwards missed the festivities that capped the conference, a gala celebration that a reporter for the *Washington Post* likened to the Academy Awards. Almost everyone in the "large-brained group" who had spent the day pondering biomedical ethics stayed into the evening to watch nine public figures receive a total of $120,000 in prize money for their work in mental retardation — as well as Waterford crystal trophies worth about another $1,000 apiece. The *Post* reporter, who usually covered society functions rather than scientific colloquia, said the trophies looked a bit like kitchen blenders.

Edwards might have enjoyed the festivities, which included his countryman Donovan singing folk tunes, Beverly Sills singing opera arias, and Joan Kennedy (Edward Kennedy's wife) playing the piano. But he was perhaps prudent to leave when he did, while his standing ovation was still ringing in the hall. His victory was destined to be brief; though he had won the audience's applause, he probably didn't change the minds of anyone on the panel or in the scientific or policy-setting community at large. And he had been called a baby killer by one of the most respected biologists in the world.

When he returned to England, Edwards was as determined as ever to proceed to human IVF. And now that Steptoe was routinely achieving success at laparoscopic retrieval of ova, such a leap was beginning to look more and more possible.

5
Fits and Starts

> Human beings, vegetables, or cosmic dust — we all dance to a mysterious tune, intoned in the distance by an invisible piper.
>
> — Albert Einstein, in the *Saturday Evening Post* (1929)

The first victory to come out of Oldham was tiny, but it made all the difference. Just before the Kennedy Foundation conference, Robert Edwards had achieved a landmark that he did not mention in Washington. Using funds and equipment borrowed from their hospitals and laboratories, squeezing their IVF work in between their more money-making responsibilities, he and Steptoe had managed to fertilize six human eggs in vitro and to grow four of them to the blastocyst stage. They began by retrieving eggs from volunteers whose ovaries had been stimulated by the fertility drugs Pergonal and HCG (human chorionic gonadotropin) and placing the eggs in a medium that had been developed by Barry Bavister, a Cambridge colleague, for culturing hamster eggs. They kept the culture dishes at room temperature before adding Edwards's sperm and then put them in an incubator. After twelve to fifteen hours, they transferred the cultures to different petri dishes, bathing them in a commercial culture medium known as Ham's F-10, to which they had added serum taken from a human and from a fetal calf.

At this point the scientists had six zygotes. They kept the pH at 7.3 (just slightly more acidic than blood), the osmotic pressure at 300 ppi (to maintain the balance of certain compounds at the level they exist at in the body), and the gases at 5 percent oxygen, 5 percent carbon dioxide, and 90 percent nitrogen. Four of the zygotes developed to the early morula stage and then, after five or six days in culture, went on to become fully formed blastocysts. The other two showed problems in the appearance of individual cells — some were "slightly mottled," with indistinct cell outlines — or in the overall shape of the morula, which was "eccentric."

But the four blastocysts that did develop — well, those four blastocysts were a wonder. "Light, transparent, floating, expanding slightly, but still smaller than a pinpoint," said Edwards, "there they were, four excellent blastocysts. The intrinsic beauty of it!" Edwards's feelings bordered on religious awe. "Like so many discoveries in science," he said, "the realization of an objective, of a goal, which clinches and clarifies years of effort makes one more humble, more aware of the wonders of Nature as they unfold in all their beauty. We had merely observed them, nothing more; someone else had designed them."

The research that Edwards and Steptoe were carrying out could not have been done in the United States — at least not with government money. It was not explicitly outlawed, but it was not encouraged, either. Any American scientists who were studying in vitro fertilization were supporting it with private money or with some clever rearrangement of grants for studies in a closely related field.

The problem with this approach was that IVF research would have benefited from the openness and coherence that federal funding would have provided. Scientists would have been forced to make their results known in a public forum, and grantees would have talked to each other and shared their experimental ups and downs. Federal funding also would have forced researchers to adhere to certain standards in order to receive their grants. But IVF was a political risk; highly organized antiabortion forces and conservative religious groups had made the issue so charged that bureaucrats did what they could to steer clear of it — even when expert advisers called in specifically to offer guidance about IVF told them that supporting such research with taxpayer money would be perfectly acceptable.

One such advisory group was the Human Embryo Research Panel, put into place in 1971. The panel, known by the acronym HERP, was asked to decide whether NIH should support studies involving human embryonic or fetal tissue and whether it was necessary — or even possible — to come up with guidelines to control their scope.

Conversations about these questions became heated, even in the collegial atmosphere of NIH. It would be unethical *not* to experiment on aborted fetuses, one scientist testified before HERP. Jerald Gaull, of the New York State Institute for Basic Research in Mental Retardation, said that such research could turn a fetus's brief existence into something meaningful. "It is a terrible perversion of ethics to throw these fetuses in the incinerator, as is usually done," Gaull said, "rather than to get some useful information."

Nonsense, said another witness, André Hellegers, director of the Kennedy Institute at Georgetown University. It's horrific, he said, to perform experiments on a doomed individual just to retrieve something useful from the tragedy. He caricatured Gaull's argument as "If it is going to die, you might as well use it." This was "the German approach," Hellegers said, an allusion to the Nazi rationale for conducting gruesome experiments on the inmates of concentration camps.

Despite Hellegers's stark imagery, HERP concluded that research on human fetuses was necessary for obtaining information to improve maternal and fetal health and that it could indeed be conducted in an ethical manner — as long as the fetus was younger than twenty weeks gestational age, lighter than 1.1 pounds, and less than 9.8 inches long, and was not kept alive artificially for more than a few hours. The panel's report went to Charles Lowe, scientific director of the National Institute for Child Health and Human Development, the institute within NIH where most fetal research would be conducted. And then the report essentially disappeared.

Politicians knew that accepting and acting on the report would provoke fury among antiabortionists — a bloc of voters that no politician, especially conservative Republicans like President Nixon and his cabinet, could risk alienating. Abortion politics was fierce during this period, made fiercer by a steady stream of state laws permitting abortion. "Right-to-lifers" did everything they could to ban the use of aborted fetuses for research, believing that allowing that use would put a positive spin on a practice they believed to be mass murder; they didn't want anyone to think that any supposed good could be extracted from an abortion, not even scientific knowledge. Other right-to-lifers did not trust scientists to conduct the experiments in a way that respected the human status of fetuses and embryos. And many were horrified at the thought of flushing fetal tissue down the laboratory sink when the experiments were finished.

On the matter of discarding excess fetuses and embryos, biology offered two counterarguments. One concerned the natural process of twinning as a contradiction of the religious question of "ensoulment" — the point at which someone is endowed with a human soul and becomes, in the view of God, a fully human being. Many religious groups taught that life begins at conception; according to this theology, ensoulment also occurs at that moment. But research had shown that up to the age of fourteen days, an embryo can split into two separate halves and become identical twins. If ensoulment occurs at conception, which twin gets the soul? How can a zygote be a human being with a single soul if it is capable of becoming two human beings? Some scientists said that pre-embryos should not be considered fully human, with fully human souls, at least until the potential for twinning had passed. According to this line of reasoning, embryos younger than fourteen days could undergo experiments that would be unethical to perform on older embryos.

As for the issue of discarding excess fetal tissue, biology had an answer to that concern, too. Even in nature, "embryo wastage" occurs on a massive scale; not every fertilized egg is destined to become a baby. Indeed, the odds are so stacked against pregnancy even in normal, young, fertile couples that it's something of a marvel that so many babies are born at all. If one hundred ova are exposed to spermatozoa through unprotected intercourse, on average sixteen will never be fertilized. Of the eighty-four that are, fifteen will be lost within a week of conception, during the preimplantation stage. Twenty-seven more will disappear by the following week during implantation, and another eleven in the next month or so. Of the one hundred ova that could be fertilized, just thirty-one develop to full-term infants, an "embryonic wastage" rate of 70 percent. Gross chromosomal abnormalities — usually resulting from problems with meiosis or from fertilization by more than one sperm — are thought to account for about one-half of these lost early embryos. The other half die off for reasons that remain a mystery.

With the natural process so prone to errors and missed opportunities, how could anyone object to the precision, the refinement, the delicate and inevitable improvement resulting from the benevolent touch of human intervention?

The abortion controversy was nothing compared to the thunderbolt that was about to strike the whole field of experimentation on humans — experimentation not just on embryos and fetuses but on any research subject who might not have really understood what he was "volunteering" for. In the summer of 1972 an article on the front page of the *New York Times* set off a scandal about a research project so outrageous, so flagrantly unethical, that its name became synonymous with the abuse of human subjects. It was the Tuskegee Syphilis Study, and the revelations infuriated politicians and theologians, medical school administrators and government bureaucrats — all of whom joined forces to set up a system that would keep such a disgrace from happening again.

The Tuskegee study began in 1932 with funding from the Public Health Service. Its goal was to record the natural history of syphilis, to see its effects on the body and their sequence when it was left untreated. Four hundred subjects — all of them male, all poor and black, all suffering from syphilis without realizing it — were enticed to join the study with offers of free medical examinations, hot meals, free aspirin, trips to the doctor, and all-expenses-paid funerals.

When the study began, syphilis was incurable. But some fifteen years later, there was indeed a cure: penicillin. That is when the ethical calculus of the experiment changed; how could you deny subjects treatment you knew to be effective? The experimental question was then moot in any event, since no one needed to understand the progression of a disease that could be completely cured. But the scientists in charge did not stop the study and did not change a single detail of the protocol. For another twenty-five years, until one of the investigators finally blew the whistle, the Tuskegee study withheld treatment from its syphilitic patients. Curing them would have interrupted the tracking of their natural downward spiral.

Sometimes the scientists had to take deliberate steps to keep subjects from getting the drugs that would have saved them. During World War II, for instance, the draft boards of about fifty participants wanted the men to get treatment for their syphilis before they enlisted. The Public Health Service persuaded the draft boards to waive the requirement, and the fifty men continued on, untreated.

Revelations about the abuses of the Tuskegee study beginning with that front-page story in the *New York Times* led to such public indignation, so many commissions and congressional hearings and follow-up reports, that it provided the first real test of the field of bioethics. The emerging discipline had no real academic home at the time, other than the Kennedy Institute at Georgetown. There was no ready source of trained experts in bioethics, no real consensus about what such training would even entail. So when Merlin DuVal, assistant secretary for health in the Department of Health, Education, and Welfare (HEW), set up the first committee to review the ethics of

Tuskegee, he had to collect a hodgepodge of experts in other fields: medicine, law, religion, labor, education, health administration, and public affairs. DuVal asked the nine-member panel to decide whether the Tuskegee study was ethical, even though no one on the committee was a trained ethicist. He also asked them what kind of regulatory action would prevent such an abuse from happening again, even though no one on the committee had much professional knowledge of public policy or political science.

The Tuskegee Syphilis Study was "ethically unjustified," the group concluded: "The withholding of penicillin, after it became generally available, amplified the injustice to which this group of human beings had already been subjected." One committee member, a Yale professor of law and psychiatry, wrote a dissent because he had wanted to use stronger language than merely "injustice." The Tuskegee subjects, he wrote, had been much more badly abused than such a mild word implied; they had been "exploited, manipulated, and deceived."

With so much public attention focused on Tuskegee, people weren't looking very closely at IVF research. Scientific developments in IVF tended to be sporadic, or stuttering, or heartbreaking, and did not sustain the popular imagination for long. For much of 1972, there was little to report anyway. Landrum Shettles almost never attended scientific meetings to give updates on his progress, and Edwards and Steptoe were encountering a maddening series of setbacks. After months of making experimental zygotes with ease, Edwards and Steptoe seemed to have lost their touch. The eggs simply stopped becoming fertilized.

Why should these tiny eggs and microscopic sperm suddenly start resisting each other? Nothing had changed, really, since a few months earlier, when fertilization of the eggs in their lab had seemed as inevitable as sex itself. The eggs were being taken at the same point in the women's cycles; the sperm were being collected under the same conditions as always; the temperature, pH, and nutrient bath were all the same . . .

Or were they? One morning in late 1972 Robert Edwards woke up thinking about that nutrient bath. For all of the previous year, he had been using Ham's F-10 — a mixture of vitamins, amino acids, sugars, and fats — to which he had added a few drops of the patient's own serum, a kind of incantation meant to mimic as closely as possible the natural environment. But because of their recent successes, the team had had to move to Kershaw's Cottage Hospital, a small private hospital two miles north of Oldham in the town of Royton. Edwards and Steptoe needed sterile facilities if they were to attempt embryo transplants, and Kershaw's offered just that — a small lab, a culture room, an anaesthesia room, bedrooms, and a bed-sitting room for the research team as well as for the patients. When the scientists moved to Kershaw's, they did so with great verve, wiping clean many of their musty habits and starting afresh with new files, new culture dishes, new solutions . . .

New solutions. Maybe that was the problem. They hadn't thought they were doing anything different, but they had failed to save their old "magic fluids," and the new batches probably differed from the old ones in some significant way. "Chemicals and drugs vary from sample to sample," Edwards observed. "It is a wise man who jealously conserves a preparation known to work." Like a World Series pitcher who refuses to shave while he's on a winning streak, Robert Edwards was superstitious about sticking with a successful formula. (When he finally started achieving pregnancies on a regular basis, he would transfer embryos only after midnight, convinced that the quietness of the lab in the wee hours of the morning had contributed to their first success and would assure success in all the others.) But in this instance, in his rush to get on with the new, he had carelessly failed to exactly follow his old procedure.

It took some months of fiddling with the solutions, but finally the Kershaw's lab was bathing its ova in a medium that was as good as the one that had worked so well at Oldham. The team fell back into the groove of fertilization. The only thing that would hold them back now was social censure. There was always some of that; the British magazine *Nova,* for instance, ran a cover story in the spring of 1972 suggesting that test tube babies were "the biggest threat since the atom bomb" and demanding that the public rein in the unpredictable scientists. "If today we do not accept the responsibility for directing the biologist," the *Nova* editors wrote, "tomorrow we may pay a bitter price — the loss of free choice and, with it, our humanity. We don't have much time left."

Such criticism was mild, however, compared to what came next. Because of some grotesque experiments by Edwards's counterparts in the United States and Sweden, a tsunami of international outrage was just about to crest.

6

Laboratory Ghouls

> I will pioneer a new way, explore unknown powers, and unfold to the world the deepest mysteries of creation.
>
> — Mary Shelley, *Frankenstein, or The Modern Prometheus* (1818)

If the scientist hadn't used the word "decapitated" to describe what he had done to the fetuses, maybe things would have turned out differently. But he was blunt about it: he had cut off the fetuses's heads, and he said so right out loud at a meeting of the American Society for Pediatric Research. Peter Adams of Case Western Reserve in Cleveland delivered his informal "poster presentation" standing next to a chart of his conclusions. Adams was just one of a horde of researchers in the exhibit hall of the San Francisco Hilton, all of whom were either making or listening to similar presentations — and his account of his work went almost unnoticed at the meeting itself. But it was picked up by a professional newspaper, *Ob/Gyn News*, in the issue of April 9, 1973. The gynecologists who received *Ob/Gyn News* for free usually glanced at it briefly and tossed it without a second thought. But not this issue, with that bombshell of a story about decapitated fetuses.

By the following day, the story was on page one of the *Washington Post*, throwing biomedical research ethics into the public spotlight in a way that echoed the previous year's focus on the Tuskegee study. The *Post* article described an investigation in which several scientists in Scandinavia, most of them Finnish except for Adams, a visiting American, took eight fetuses that had been aborted by hysterotomy (opening up of the uterus) at the gestational age of twelve to seventeen weeks, kept them alive just long enough to keep the blood flowing to their brains, decapitated them, and attached the severed fetal heads to an apparatus that provided sufficient oxygen and nutrients to keep the brains functioning. The goal was to measure the rate of perfusion of metabolites that had not previously been thought to cross the placental barrier.

The purpose of the experiment was to determine whether glucose or D-beta hydroxybutyrate was the more important source of energy in brain development in utero. But the image of a laboratory filled with severed fetal heads was more than most people could bear. "If there is a more unspeakable crime than abortion itself," said Cardinal John Krol of the Philadelphia archdiocese, "it is using victims of abortion as living human guinea pigs."

The day after the *Post* article appeared, two hundred students from a prominent Catholic girls' school in Maryland, Stone Ridge Country Day School of the Sacred Heart, marched off their bucolic campus in Bethesda, where the bright yellow forsythia blossoms were at their peak. The girls headed across the traffic of Rockville Pike and over to the National Institutes of Health, where they held an impromptu teach-in in an auditorium. Three students led the group: Renee Meier, Theo Tuomey, and Maria Shriver. Seventeen-year-old Maria had learned to be indignant about fetal research at home; her mother was Eunice Shriver, who had organized the 1971 bioethics conference in Washington at which a Nobel laureate had called Robert Edwards, the world's most eminent IVF researcher, a baby killer.

As a sign of the clout of young Maria Shriver and her classmates, the NIH deputy director for science showed up at the Stone Ridge teach-in. He stated flatly that the federal government did not sponsor any research involving living aborted fetuses, and he promised not to support any such studies in the future. But the girls had all the cynicism of 1970s adolescents. "Why are they drawing up guidelines," asked one, "if they don't intend to use live fetuses?"

The NIH officials were drawing up guidelines because that was, in essence, what bureaucrats did. Guidelines were one of the few methods available to the federal government — along with regulations, commissions, moratoriums, and hearings — to deal with thorny issues. Sometimes, however, thorny issues did not fit neatly into an ordinary regulatory scheme. If fetal research was banned, for instance, did that mean embryo research was banned, too? And if embryo research was banned, was IVF also banned, since it involved creating embryos? Is a preimplantation embryo morally equivalent to a postabortion fetus? Is it relevant that one is created for the purpose of implantation and eventual birth, while the other is destined to die?

Questions like these were posed to a panel of twenty-three experts assembled in mid-1973 to help NIH decide whether the government should fund IVF research. The committee comprised prominent physicians, lawyers, public interest advocates, and one theologian. More panel members than on previous panels identified themselves as ethicists, an indication that the field of bioethics was beginning to take shape. Leon Kass, then a scholar at the Hastings Center, was on the panel, as was the center's vice president, Daniel Singer, and LeRoy Walters, the new director of the Center for Bioethics at Georgetown, part of the Kennedy Institute for Bioethics. The group included four professors from Harvard: a professor of population ethics, the chairman of the pediatrics department, a professor of education and social psychology, and the director of the Center for Evaluation of Clinical Procedures. There were pediatricians from Johns Hopkins, Columbia, and Yale; representatives of the Children's Defense Fund and the Center for Law and Social Policy; a lawyer and an obstetrician from the University of Pennsylvania; and a professor of social ethics from Union Theological Seminary. And there was Raymond Vande Wiele, chairman of the department of obstetrics and gynecology at Columbia Presbyterian Hospital.

The panelists agreed that the federal government should fund in vitro fertilization studies — though, like its predecessors, this committee urged that certain safeguards be put in place to guarantee that the work was done ethically. Until those guidelines were written, the panel recommended that "all experimentation on extracorporeal fertilization of human eggs shall not be further supported at the present time."

Strangely, however, one influential government scientist, William Sadler, seemed to think such research was still allowed — a significant contradiction, since Sadler, as the chief of NIH's population and reproduction grants branch, would have overseen any NIH grants involving human IVF. In the fall of 1973 Landrum Shettles wrote to Sadler, asking him to spell out the agency's position. "To the best of my knowledge," Sadler wrote back, "support for [IVF] studies is not proscribed by law or by official NIH policy." The fact that knowledgeable men could disagree so completely on whether the government would or would not support research on in vitro fertilization was testament to the confusion surrounding the issue — confusion that would persist for years to come.

First the revelations about the Tuskegee study abuses; then, just ten months later, the account of grisly research on fetuses. Among the public, the suspicion grew in the spring of 1973 that if left to their own devices, at least some scientists would go well beyond the bounds of decency. When people wonder what scientists are doing without their knowledge, it is time to bring the public into the decision-making process, through congressional hearings, open meetings of advisory commissions, and requests for comments on proposed regulations. These are the ways that nonscientists serve as watchdogs and let the scientists know when they have gone too far. Usually only the most committed, or the most repelled, observers become actively involved, but the mechanisms exist for anyone to make a statement.

Politicians can serve as watchdogs, too, by introducing legislation that would impose certain standards or restrictions on scientific research. In mid-1973, Senator Edward Kennedy, chairman of the

Senate Health Subcommittee and one of the most influential men on Capitol Hill, introduced a bill creating a National Commission for the Protection of Human Subjects of Biomedical and Behavioral Research. Kennedy, reflecting in part the heightened sensitivities of his sister Eunice, was especially interested in protecting those who could not give voluntary and fully informed consent on their own: prisoners, patients in psychiatric institutions, the mentally retarded, children, and fetuses. Either because a potential for coercion was implicit in their living situations or because their limited capacities made them unable to understand benefits and risks, such subjects required more protection than anyone else.

Representatives of the Nixon administration testified against the Kennedy bill in July. They preferred to maintain the oversight arrangements already in place at the hospital and university level, which specified that all human subject research conducted under government grants must be approved by an institutional "ethical review committee." Conservatives, including Senator James L. Buckley of New York, made their support of the Kennedy bill contingent on the inclusion of a permanent ban on all fetal research. As a compromise, Kennedy added an amendment that would place a temporary ban on fetal research, pending an ethical review by the national commission. The amendment said nothing about IVF specifically, but no one in the government wanted to support research on human embryos while research on fetuses was prohibited. The bill that was finally signed into law in 1974 required that the commission present a report on fetal research to the secretary of HEW within four months — at which time, if the commission recommended it, the ban could be lifted.

Laws and regulations are not the only ways to control a new technology. Sometimes, a scientific development takes place because someone can make a profit from it. That is what happened with in vitro fertilization. Market forces helped determine the site of the first IVF clinic in America. Instead of being in New York or Boston or Chicago or Los Angeles, where you'd expect to find it, the first clinic opened at the Eastern Virginia Medical School in Norfolk, Virginia.

Norfolk was a faded port city, a sleepy southern town as famous for its fragrant crape myrtles as for its rough edges of peep shows and liquor stores. In the sixties it began to suffer the urban collapse that plagued many cities with populations of less than one million; the unsavory downtown street life kept middle-class people away at night, driving them to spend their leisure dollars far from the center of the city. This decline led to a massive urban renewal project, the first such project in the South. The plan contained an unprecedented element: the construction of a medical school to serve as the centerpiece of a rejuvenated downtown. The school was the brainchild of Mason Andrews, a Norfolk native, a fixture on the city council, and, as a busy obstetrician, someone who had been present at the birth of a healthy percentage of Norfolkians. The only time Andrews had lived outside the city limits was from 1936 to 1947, when he was an undergraduate at Princeton and, later, a medical student, intern, and resident in obstetrics and gynecology at Johns Hopkins. As soon as he completed his training, he headed back home to Norfolk to go into private practice.

As a member of the city council, Andrews pushed for the new medical school. The idea was risky. Two excellent medical schools were nearby, the University of Virginia in Charlottesville, a three-hour drive west along Interstate 64, and the Medical College of Virginia at Richmond, just ninety minutes to the south. But the Norfolk city fathers were willing to give it a try.

At first it looked like a mistake. For years after the Eastern Virginia Medical School opened, Andrews watched his creation languish in the shadow of the two more prestigious schools. He worried that no amount of funding from the city budget, bolstered by grants from the Andrews family estate, would keep the medical school from floundering. Something dramatic, he knew, had to be done.

Part Two

The Modern Prometheus

7
Toward Happily Ever After

> The individual organism is transient, but its embryonic substance, which produces the mortal tissues, preserves itself, imperishable, everlasting, and constant.
>
> — William Osler, "Counsels and Ideals" (1906)

The test tube containing the Del-Zio gametes that sat on Raymond Vande Wiele's coffee table seemed almost electrically charged. Its very existence was disquieting, a perturbation in Vande Wiele's world of order and procedure. He worried that Shettles hadn't gone through proper channels to get the IVF effort approved by the hospital committee in charge of human experimentation. He worried that such an oversight, flouting the letter of assurance that Columbia had signed with funding agencies, would endanger the status of all federal grants for research at the university. He worried especially about the ethics of the procedure. What if the fetus was damaged, developmentally or chromosomally, by its unusual start in life? Whose responsibility would that be? And who did Shettles think he was, anyway, forging ahead on his scientific quest without seeming to care about the human consequences? It wouldn't be Shettles who would take home the mistakes; it would be the baby's parents. How could he have so little regard for that essential truth?

It wasn't that Vande Wiele was a diehard opponent of in vitro fertilization. Even though he had been raised a Catholic, trained at Catholic University in Louvain, Belgium, he did not object to IVF on the traditional Catholic grounds that it was unwarranted intrusion into God's terrain. A scientist above all else, and a longtime convert to the Episcopal Church, he thought IVF offered a chance to unwrap the dark secrets of fertilization and early embryonic development. Bringing this recondite process out into the open in the laboratory could, he believed, help explain the unknowns of chromosomal replication, cell division, cell differentiation — knowledge that could, in turn, offer insights into birth defects, contraception, and cancer.

But Vande Wiele believed IVF had to be done by the right individuals following the right procedures at the right time. Shettles, he believed, was most definitely not the right individual. Robert Edwards, whom Vande Wiele knew from international meetings he had attended, might have been the right individual, but he was not, in Vande Wiele's opinion, waiting for the right time. For years Vande Wiele had criticized the British investigator's speed in moving from external fertilization and embryo transfer in animals to the same manipulations using human eggs and sperm. Too soon, he kept saying; it's unethical to try this trick yet with human beings. No one could say with confidence that IVF was safe. Vande Wiele thought it absolutely necessary to know that IVF could be done safely in other species before messing around with human gametes. And now here was someone in his own department, trying to do the same procedure as Edwards, just as prematurely — and, even worse, in an incompetent, irresponsible manner — right under his nose.

The chairman made a few more phone calls, then paged Shettles on his beeper. By now it was midmorning on Thursday, September 13, 1973. Come to my office, he said, trying to keep his voice noncommittal, at two o'clock this afternoon.

The Del-Zio test tube was still in the beaker on Vande Wiele's coffee table when Shettles arrived. The stopper had been removed, and the mixture had been at room temperature for hours. In all likelihood, cell division had already ceased. Next to the open test tube was a tape recorder, switched to "record." And sitting beside Vande Wiele was

Duane Todd, associate chief of staff at Presbyterian Hospital. Clearly this was not going to be just a friendly chat with the chairman.

"Dr. Allerand brought this to me," Vande Wiele said with deliberate calm, never stating that it had been his idea, not Allerand's, to take the test tube out of the incubator in the first place. "Now, could you tell me what it is?"

Shettles answered in his high-pitched Truman Capote whir, speaking so rapidly he was sometimes difficult to understand. "That, that's some, uh, fluid and, and uh, ova, and a bit of tubal mucosa that Dr. Sweeney did, he did a, a laparotomy . . . I don't know if the ova is in our follicular fluid . . . then there's tubal mucosa, uh, and, and the husband's sperm, and when everything's settled, uh, serum, some serum is incubated with, uh, atmospheric oxygen . . . We want to see if it looks like, if it looks like it's going to be normal, then I guess try to, try to place, place the ovum, it would be a blastocyst, or a blastocyst more or less, but in utero . . . see, see if it would take. That's what we wanted to do."

Vande Wiele, his accent cultured and urbane, was more glib. Indeed, as he laid out his argument about why he felt compelled to stop the experiment, he sounded a bit like Agatha Christie's immortal Belgian detective, the debonair and highly rational Hercule Poirot. He told Shettles that he believed this kind of research was prohibited by the government's main funding agency for biomedical research, the National Institutes of Health, and that to allow it to proceed would endanger all of the federal grants awarded to Columbia University scientists at the time. He said he thought the procedure was sloppy — "you don't have this sterile, as you well know" — and to his mind unethical.

"Landie, Landie," Vande Wiele sighed at one point. "Let me, let me ask you a few things. I mean, aren't you concerned at all that if you would implant a baby like this, that the baby would be malformed? . . . [If] this were an abnormal child, how would you be able to live with yourself?"

Shettles didn't answer directly, so Vande Wiele read to him from an article in the *New England Journal of Medicine.* It said that while Edwards and Steptoe were proceeding cautiously and would not implant an IVF embryo until they could "rule out the presence of genetic or other defects," here in the United States Landrum Shettles was, "in contrast," ready to proceed, consequences be damned. "I mean," said Vande Wiele, "what, what kind of, of, of a, of a position does it put not only this department but the whole medical school in?" He was starting to lose his composure. "I mean, here on one side, the English investigators are concerned about ethical problems, and Dr. Shettles is not."

Vande Wiele concluded by saying that he would dispose of the test tube. Shettles offered no protest. But on his way out of the office, as he passed the table where the test tube sat, Shettles said casually, as if the thought had just occurred to him, "Let me just throw this thing out, okay?"

Vande Wiele would not be caught off-guard. "Leave it here, no, please," he said. Only the abruptness of his clipped response betrayed the rage he was trying to hide.

Like many of his colleagues, Vande Wiele worried about human IVF on moral as well as scientific grounds, as he tried to get across to Shettles in their tense meeting that afternoon. But Shettles didn't think about how he'd feel if "the baby would be malformed" — chiefly because he didn't expect that it *would* be malformed. He had an almost mystical belief in the inherent wisdom of the zygote. Rather than fearing the damaged fruits of IVF, he trusted the natural process of miscarriage to cast out any embryos with mental or physical defects created by his lab manipulations. He simply never gave serious thought to the possibility that he could be contributing to the birth of grotesques.

This sanguine approach to the what-if's of his experimentation put Shettles in direct opposition to the bioethicists, whose professional role it was to think through the what-if's of every advancement, to tease out the hidden dangers of developments that seemed on the surface to be pure progress. Some of their declarations were full of fire and brimstone, reflecting the religious background of so many of them. "Any man's or any woman's venture to begin human life in this way is morally forbidden," said Paul Ramsey, the Princeton professor of religious studies, in a typical comment. "We cannot morally get to know how to perfect this technique to relieve human infertility," nor should we even try, since IVF would be nothing less than "a disastrous further step toward the evil design of manufacturing our posterity."

But not every bioethicist was opposed to IVF. One of the most prominent, John Fletcher of the University of Virginia, was one of IVF's staunchest defenders — though not necessarily for the reasons that scientists wanted it defended. Fletcher had come to bioethics the way many of its pioneers did, through theology; he started out as an Episcopal priest. He was also the founder of a school of thought called "situational ethics," which said that there is no objective morality, neither absolute good nor absolute evil. The only thing that is purely good, said Fletcher, is love, and "all laws and rules and principles and ideals and norms are only contingent, only valid if they happen to serve love in any situation."

Applying this philosophy to IVF, Fletcher concluded that Shettles's efforts were moral because they made life better, because they maximized human good and minimized human harm. Yes, IVF had some frightening implications, Fletcher admitted, and would no doubt send scientists down a slippery slope that allowed them to manipulate human destiny in dozens of inappropriate ways. But the slipperiness didn't trouble him; there would be time enough to apply the brakes as soon as a downward spiral became clear. The alternative — to shut the door on experimentation even before it led to anything — was worse, he believed. It was willfully to choose ignorance, and "dangerous ignorance is far more dangerous than dangerous knowledge."

The truly arrogant act, said Fletcher, would be the failure to try IVF. "To deny patients their initiative and choice in the matter is to impose on them one's own subjective theory of prenatal life." He considered a man like Shettles to be "the modern Prometheus," proud bearer of a name derived from the Greek word for "forethought," a visionary who could "think ahead to emerging needs and options."

What an infelicitous turn of phrase. "Modern Prometheus" was also the subtitle of Mary Shelley's classic work *Frankenstein.* Dr. Frankenstein's monster — or, more accurately, its cinematic incarnation, as played by Boris Karloff — was already all mixed up, in the public imagination, with test tube babies. And here John Fletcher was unwittingly stressing that link. If Shettles was a "modern Prometheus," then perhaps he, like Dr. Frankenstein, was driven by a hubris blind to its own awful consequences. Did the public dare let Shettles's experiments continue without restraint in the twentieth century, knowing what had been unleashed by his fictional counterpart in the nineteenth?

By the afternoon of September 13, the test tube containing the Del-Zio eggs and sperm was in a deep freeze at Columbia University. Vande Wiele said he had it placed there to preserve the gametes, though he never said why he wanted them preserved. But preserved or not, the gametes almost certainly were no longer alive; even if they had survived their prolonged exposure to room temperature, the abrupt freezing would have killed them.

Doris Del-Zio was still in post-operative pain in her room at New York Hospital. She had anticipated the pain; in all her previous attempts at infertility treatment, she had been forced to forego post-surgical pain relief, since she seemed to be allergic to just about every analgesic on the shelf. But she faced it bravely. She was "happy and hopeful," as she would later recall; "it was a happy pain."

At about four o'clock John Del-Zio was called away from his wife's bedside to take a phone call at the nursing station from Doris's surgeon, William Sweeney.

"The procedure has been terminated," Sweeney said. John barely understood what he was hearing. "A doctor over at Presbyterian, his name is Dr. Vande Wiele, he destroyed the culture. He says there's an NIH ban against this kind of procedure." The line was silent at John's end.

"I'm sorry, John," Sweeney continued. "I know this is a dastardly time to have to tell Doris."

John decided to keep the full truth from his wife until the next morning, when she might feel stronger. For now he wanted to alert her to the possibility that things might not be working out the way they had planned.

"That was Dr. Sweeney," he said when he got back to Doris's room. "He called to tell us the eggs look like they're not fertilizing."

Doris was unfazed. "I just know that they've taken," she said. "They need hours and hours to fertilize. Don't worry."

But John had trouble putting the best face on things, and he left soon afterward for his sister's house in Queens, where he and Tammy, Doris's ten-year-old daughter, were staying.

Later that evening, visiting hours over, Doris got a phone call of her own. It was about nine or nine-thirty, and the caller was a distraught Landrum Shettles.

"I'm so so sorry for what happened," he began in his southern twang. He had never met Doris and had not spoken to her.

"What do you mean?" asked Doris. "What are you sorry about?"

"Didn't Dr. Sweeney talk to you?"

No, Doris said, and told Shettles about the phone call at the nurses' station. "He told my husband that it didn't look like the eggs had fertilized yet," she said, "but I'm sure everything is okay . . ."

No, everything is not okay, Shettles interrupted her. Another man, a Dr. Vande Wiele, destroyed the culture.

"What? What are you saying?" Doris asked, confused. "Why would he do a thing like that?"

"He believed there was a ban on this type of research. He thought the National Institutes of Health didn't allow it. So he opened up the incubator."

"Are you sure?" Doris asked. "Are you sure there wasn't just some sort of mistake?"

Shettles apologized to Doris — for telling her in this way, for upsetting her, for all the years of pain and surgery that had come to nothing, had come to this final bitter disappointment. He kept telling her how sorry, how very sorry, he was. And Doris just kept asking, "Are you sure? Are you sure?"

The next morning, September 14, Dominique Toran-Allerand opened the door to her laboratory and strangled a yelp. Her microscope equipment was missing. The carbon dioxide tanks attached to the incubator had been opened and emptied. The sink drains had been stuffed shut and all the faucets turned on — Allerand was standing ankle-deep in water. She had no doubt who was responsible for this mayhem. Only one person other than Allerand and Dr. Parshley had a key to the laboratory: Landrum Shettles.

Not only did Shettles deny that he was responsible, he said that *he* was the vandal's intended target. He cited the destruction of the lab — the lab, after all, where *his* test tube had been stored — as proof that someone at the university was out to get him.

Santiago Ramon y Cajal had an explanation for persecution like this, Shettles told himself. The Spanish neuropathologist, who won the Nobel Prize in 1906, once wrote a sentence about scientific pioneers that Shettles had taken earnestly to heart, copying the quote, posting it on his bulletin board, handing out copies to his children when they visited him in his cramped quarters at Columbia Presbyterian. "To be right before the right time is heresy," Cajal wrote, "sometimes to be paid for with martyrdom."

Vande Wiele ordered an emergency meeting of his department's executive committee, and on Monday, September 17, 1973, they voted unanimously to ask for Shettles's resignation. Shettles wrote the requested letter one month later, pointing out that he would have liked to stay on were it not for "certain irreconcilable differences which have arisen between myself and certain hospital personnel." By October 17 his official relationship with Columbia was over. He was offered six months' salary as severance pay, but not the pension he would have received had he remained at Columbia until the age of sixty-five — just thirteen months more.

Despite the firing, Shettles continued to hang around the medical school. He was still spotted gliding through hallways where he didn't belong, still thought to be living in the hospital as he always had, a rootless husband and father without any other real home.

By November Mary Parshley, who had lent Shettles the use of "her" incubator, finally realized there was no room for her in the sixteenth-floor laboratory she was supposed to have given up the previous July. She turned the lab over for good to her successor and formally retired.

At last Dominique Toran-Allerand had the lab to herself. The first thing she did was change the lock.

8

Baby Dreams

> Is there possibly some wisdom in that mystery of nature which joins the pleasure of sex, the communication of love, and the desire for children in the very activity by which we continue the chain of human existence?
>
> — Paul Ramsey, "Shall We 'Reproduce'?" (1972)

The longing for a child is so entangled with all other human longings that it sometimes seems as inevitable as breath. Even the author of *Frankenstein* herself, whose fiction danced along the fuzzy borders between life and not-life, between creation and devastation, knew what it was like to ache with baby lust. Mary Shelley, born Mary Godwin, lived under the cloud of tragic childbirth almost from her first moments on earth; her mother died ten days after giving birth to her. And then in early 1815 she suffered her own tragedy, watching her prematurely born daughter die within days. The baby had been conceived when Godwin was barely seventeen years old and living in a rented cottage with her soulmate, Percy Bysshe Shelley, who at the time was still married. Death and birth were thus tragically intertwined for Mary Godwin Shelley, and she had a vivid dream about her lost child. "Dream that my little baby came to life again," she wrote in her journal a few days after her daughter died; "that it had only been cold, and that we rubbed it before the fire, and it lived." Her archetypal dream, one that had haunted grieving mothers for centuries and would continue to do so, proved prescient. Its leitmotif was similar to that of another dream several months later, a dream that inspired Mary Shelley's novel about Victor Frankenstein, who made his own creation "come to life" by rubbing it with a warm, firelike current of electricity.

And the dream of a dead infant come back to life echoed in the nightmares that Doris Del-Zio reported after her failed attempt at IVF.

In the fall of 1973, Doris said, her sleep was plagued by sad, disturbing dreams, especially a recurring one about an empty baby blanket. In the dream, she heard a baby crying and saw that it was wrapped in a soft blanket patterned with tiny animals. She ran over to soothe the baby but found the blanket open and nothing inside. Still she heard a baby's cry, and she would wake up crying herself, frustrated and horrified by her inability to reach it.

It was an awkward time in history to be haunted by such deep longings for a child. Against this powerful, instinctual pull, one segment of society was pulling just as hard in the opposite direction. While Doris was so zealously pursuing the traditional roles of mother and wife — and, ironically, feeling so devastated by her failure to achieve motherhood that she was abrogating the responsibilities of housekeeper, cook, confidante, and lover — the emerging social movements of the early 1970s were undermining those very roles. Leftists said that it was a sorry thing to bring children into this world; environmentalists said that more babies would further despoil the planet; feminists said that marriage as an institution was sexist and suffocating.

The arguments against traditional motherhood were loud and hard to ignore. As Doris lay in her bed at New York Hospital, recovering from yet another tubal surgery and from the miscarriage of her IVF attempt, it would have been impossible not to overhear the raucous "Battle of the Sexes" tennis match on the corridor TV. Just eight days after her futile operation came the much-anticipated showdown between Billie Jean King, at twenty-nine one of the most famous athletes in America, and Bobby Riggs, a fifty-five-year-old one-time Wimbledon champion turned self-described hustler and con man.

The publicity that heralded the match (watched by 50 million people) presented the showdown as a renunciation of the roles that Doris most deeply desired.

Billie Jean King was carried out to the Houston Astrodome in a golden litter such as Cleopatra might have used, held high by four muscular young men dressed like Egyptian slaves. Bobby Riggs, surprisingly self-deprecating for all the bravado that had preceded the September 20 face-off, arrived in a rickshaw pulled by a bevy of women called, aptly, "Bobby's Bosom Buddies."

King wiped up her opponent in three sets, six-four, six-three, six-three, in what was called "the drop shot and volley heard 'round the world." Poor Bobby Riggs, horn-rimmed spectacles sliding down his sweat-slicked nose as he raced after King's winning lobs, never had a chance.

Like most of the women watching the match that Thursday night, the nurses on Doris's floor at New York Hospital were surely cheering for Billie Jean, seeing her as proof that a woman could be every bit as tough as a man — a man twenty-five years older than she, yes, but a man who had made a great show of insisting that no woman, no matter what her age, could ever beat him. Billie Jean's victory was a chink in the dam for women who refused to be hemmed in by social expectations. The Battle of the Sexes wasn't playing out just in the professional sports arena in 1973; it was taking place in day-to-day workplaces, too.

Many of the feminist struggles were largely symbolic, but, like so much that happened during this time, the symbols took on a charged meaning that made the struggles seem all the more valiant and worth fighting for. This seemed especially true in health care. In Chicago, for instance, the head nurse in the cardiac care unit of Northwestern Memorial Hospital sued when she was fired for refusing to wear one of the perky little white caps that female nurses — but not male nurses — were required to wear at all times. She saw the cap as a symbol. "As cardiology nurses, we made split-second, life-and-death decisions for our patients," she said. "And yet the nursing administration would not let us make decisions about the way we dressed."

Other struggles in the health care professions were more substantive, leading, among other things, to a steadily growing number of women choosing to become doctors. In 1970 only 7 percent of physicians were women, and women made up 9 percent of the freshman class in medical schools. By 1990 women doctors were 17 percent of the total, and by 1999, 46 percent. This change in medical personnel had profound implications for the treatment of infertility. Among new medical school graduates, a larger proportion of women than of men went into obstetrics and gynecology. And ob-gyn departments were the source of medical brainpower for the new subspecialty of infertility — in some schools called maternal and fetal health — which took as its mission making babies for couples who were having trouble making them on their own.

Feminism was a growing movement, but it was clearly not for everyone; just look at the sharp contrast between the aspirations of Billie Jean King and those of Doris Del-Zio, who were exact contemporaries (both were born in November 1943, just nine days apart). Millions of women, who had no interest in reading *The Female Eunuch* or *The Second Sex,* wanted nothing more out of life than to stay home and raise their kids. These were the women who disdained *Fear of Flying,* Erica Jong's controversial feminist novel of 1973, with its famous paeans to the joys of casual, anonymous sex, and instead read *The Total Woman,* an antifeminist how-to book written by a Florida housewife named Marabel Morgan. To Morgan, maintaining a loving, monogamous marriage was the loftiest goal of the Western world, and to meet that goal a wife should focus not on her own needs but on her husband's, should treat him like the king of the castle, should subvert her own desires and stay home to cook meatloaf and greet her man after his long hard day of breadwinning, with his martini in hand and wearing nothing but a homemade Saran Wrap bikini.

Except perhaps for the Saran Wrap, Doris Del-Zio's dreams and aspirations put her squarely in the Total Woman camp.

Another social movement of the seventies preached against the instinct to have children: Zero Population Growth (ZPG) embodied the rallying cry that "the personal is political." The movement was started by Paul Ehrlich, a butterfly biologist at Stanford, who wrote *The Population Bomb*, a forceful defense of planet Earth in the tradition of Rachel Carson's *Silent Spring*. Ehrlich took up as his life's work the dissemination of its message — that the single biggest threat to our global environment was overpopulation. For believers in ZPG, the most intimate of decisions could be interpreted as a declaration of one's commitment to the environment.

ZPG urged people to tread softly on the earth, to conserve resources, and to ensure a sustainable world in the future by limiting the size of their families. Its slogan was clear and direct: "Stop at two." And largely because of its clarity, the message was incredibly effective. In the early 1960s the fertility rate in the United States was 3.4 children per woman, with the population growing at a rate of 1 percent a year. By 1975, taking heed of the "stop at two" campaign — and responding as well to other economic and social trends — Americans were reproducing at the rate of only 1.8 children per woman, and the overall population was actually declining each year by 1 percent.

Even with all these competing pressures, however, the instinctual drive to have children remained so powerful that it all but guaranteed a market for anyone who could help people have babies of their own. It was estimated that 7 percent of married couples were infertile — meaning that they had tried to conceive, without success, for at least one year. Some 4 million American couples, and no doubt millions more worldwide, might do almost anything, pay almost any price, to have the children they yearned for — couples who would have been delighted to stop at two, or even at one, if the babies were their own.

In the fall of 1973 Doris Del-Zio was not quite thirty years old, pretty and plump, her dark hair in a bouffant flip that was as unstylish as her attitude toward matrimony. She looked like a rounder Lesley Gore, whose sixties hit "It's My Party and I'll Cry If I Want To" seemed a perfect accompaniment for Doris's perennial pout. She had been a happy-go-lucky young woman, she liked to say, but her experience with infertility had left her bitter and withdrawn.

Doris had had more than her share of medical mishaps: a bout of polio at age nine had left her paralyzed for a year and with residual weakness in her legs; her tonsils, appendix, and gallbladder had been removed by the time she reached her mid-twenties; she had lifelong asthma; and she had undergone a string of gynecological operations, all connected to her five-year struggle to have a baby, to open her tubes or remove ovarian cysts. But she said it was the events of September 13, 1973, when the test tube filled with a mixture of her ovarian tissue and her husband's sperm was put into a deep freeze, that threw her into a state of despair from which she expected never to recover.

She hardly ever went outside except to pick up Tammy after school. She gained weight, turning frumpy and matronly, and stopped inviting friends in or going out socially. She had been, her husband said, a "happy girl," but suddenly she was gloomy, self-deprecating, and always sad.

While high-tech science was failing Doris Del-Zio, it was captivating television audiences with shows like *The Six Million Dollar Man,* which in late November 1973 began what would be a successful five-season run. The show's hero, an air force colonel named Steve Austin, played by Lee Majors (husband of the *Charlie's Angels* bombshell Farah Fawcett), was one of the new, bizarre sci-fi hybrids known as a cyborg — part man, part machine. The scriptwriters called him "bionic."

The hybrid nature of the Bionic Man — everything about him ordinary (if preternaturally handsome) except his manmade, nuclear-powered legs, right arm, and left eye — reflected the two-sided image of science itself in the early seventies. Science was awesome, exhilarating, and definitely something you wanted to have on your side. But it was also unnatural and frightening, a territory better left unexplored. Who knew what dark powers such meddling could unleash?

Christmas of 1973 brought little joy to the Del-Zio household. John and Doris marked their fifth anniversary on December 20, three months after their IVF misadventure, and a letter that Doris inserted into her anniversary card to John made it clear that she continued to long for the baby they couldn't have. "It's been five years for us, for the most part it's been five difficult years," she wrote on pale blue deckle-edged stationery, the kind girls once used to write letters home from camp. Like the paper itself, the handwriting was childish, a loopy, back-sloping script with lots of fancy swirls.

> I'm sorry I've failed to give you the child you wanted, I really tried and still am. But, I'm in a vacuum now, the only person I can feel love from and give it back to, is Tammy. Maybe it's because I feel safe with her love, I don't know. I want to be truthful with you, right now I feel nothing with us, I can't pretend. Too much has happened to me in the last few months, somehow I have to bring myself back to normal, I don't know how. So far, you haven't been able to help me, I don't know who can help, but I have to find help somewhere. It scares me. I don't like being in this vacuum.
>
> If you feel this is going to be too much for you to handle, let me know. I'll take Tam and go somewhere alone. If you decide it's not going to be a burden to you then remember, a lot will be

falling on your shoulders, such as your mother. . . . you'll have to tell her I don't want to be bothered and keep me away as much as possible. I need to be alone, help me if you can, I need it.

Sorry this anniversary isn't a happy one for you, it's all my fault and I can't help it.

Very little seemed to change for the Del-Zios as 1973 blurred into 1974. In February Doris returned to New York to have Sweeney operate on her yet again, this time to remove a mass in her uterus. Her Florida gynecologist had diagnosed the mass as an early pregnancy, but Doris knew that was impossible. She was so humiliated by the failure of her attempt at IVF, she said, that she "couldn't look at my husband, much less have sex with him." The last time she and John had made love was on September 10, 1973 — the day before her hospital admission for IVF, and John's fifty-sixth birthday.

Fetal research made for some anachronistic police work in April 1974, when a 160-year-old Massachusetts law against grave robbing was invoked in the arrest of four Boston physicians. The doctors had been studying fetal metabolism of drugs — specifically, whether antibiotics crossed the placental barrier, the first step in judging whether such drugs could be used prenatally to prevent congenital syphilis, a devastating and sometimes fatal disease. The doctors found pregnant women who were planning to have abortions, gave them antibiotics for a few days, and took tissue samples from the fetuses after the abortions to look for traces of the drugs. But the mothers, even though they had agreed to take the drugs before the abortion, hadn't given consent for what amounted to a postabortion autopsy. The doctors were charged with illegally "removing and conveying away" human bodies.

At the time the bill creating the National Commission for the Protection of Human Subjects of Biomedical and Behavioral Research, with its four-month ban on fetal tissue research, had not yet been signed into law. When the Boston physicians conducted their research on aborted fetuses, the federal government had taken no formal position restricting fetal research, had imposed no ban, passed no legislation, issued no regulations. So the Massachusetts authorities had to exhume an 1814 law, broadly interpreted and creatively applied, to try to put a stop to it.

In June 1974, Doris Del-Zio said, she was shopping in a Florida department store when she suddenly passed out and fell to the floor. When she came to, a concerned salesgirl was asking if she was all right. Doris looked down and saw that her arms were filled with baby clothes.

June was the month when, if everything had gone as planned, the Del-Zios' test tube baby would have been born.

One month later, in vitro fertilization was in the headlines with the stunning news that three toddlers conceived through IVF were alive and well somewhere in Europe. The news was issued, almost as an afterthought, by a soft-spoken professor of gynecology from England. Douglas C. A. Bevis, fifty-five, had recently been appointed assistant chairman of obstetrics and gynecology at the University of Leeds. Usually he had nothing much to say in response to journalists' questions about the progress of his IVF efforts, which had been going on just about as long as those of Edwards and Steptoe. But this time he threw back a stunning rejoinder. It was the style of his announcement, almost as much as its content, that was so surprising; one magazine described it as "quite uncharacteristic — and totally un-British." It came on Monday, July 15, 1974, during the first day of the annual meeting of the British Medical Association, held that year at the University of Hull, two hundred miles north of London. Bevis was making his way to the lecture theater to deliver a talk titled "Limits of Interference in Medicine." As he was about to enter, a few reporters blocked his path and asked whether anyone had had any success with IVF that year. Yes, Bevis said, much to the reporters' shock, IVF had indeed worked. There were now three babies alive as a result of in vitro fertilization.

"A matter of luck," was all he said by way of explanation, offering no further details: no names, no cities, no countries, not even the name of the scientist involved other than that he was a gynecologist whom Bevis knew. "So many have been attempted," he said, "that by the law of averages some have come through." The only specifics he gave were that one baby lived in England and two on the Continent and that they ranged in age from a year to a year and a half — meaning that the first had been born right around the time that Shettles and Vande Wiele were having their confrontation on the other side of the Atlantic.

In his formal remarks to the British Medical Association, Bevis made no mention of these mysterious IVF successes. But by the time his lecture was over, the British public had already heard about it, through news flashes on radio and TV. Bevis was forced — almost against his will, it seemed — to hold a press conference.

As further details came out, a reporter for the American publication *Medical World News* noted that the particulars "coincided closely" with those offered by Edwards and Steptoe in their reports of their own successful embryo implants. Bevis described all the steps that Edwards and Steptoe had described: fertility drugs to induce superovulation, removal of ovarian tissue, harvesting of eggs, fertilization with the husband's sperm, incubation for five to seven days, implantation through the vagina and cervix directly into the uterus. But when reporters called Steptoe for a comment, he said the IVF success wasn't his — and he was angry at Bevis for suggesting that it might be. "I am astounded that Professor Bevis should have made this statement," Steptoe said. "As far as I know no one in this country or anywhere else has yet succeeded in this technique."

By the next day Bevis was trying to extricate himself from what one scientist was calling his "bed of nettles." A friend from Leeds came to his defense, saying Bevis had made his announcement not to claim credit but merely to show that IVF did not inevitably lead to chromosomal damage, since the three children all seemed to be healthy and normal. The friend said it was imperative to maintain the anonymity of everyone involved, families and researchers alike, to protect the children's identities. It would be "disastrous," he said, for the public ever to know these test tube toddlers' names.

The following day, Wednesday, July 17, Bevis suddenly decided to get specific. He told the medical correspondent for the London *Daily Mail* that he was the scientist who had created the three IVF babies. Two were living in England, he said, changing his original story slightly, and the third in Italy. Bevis said he had worked on their conception essentially on his own while at the University of Sheffield. He had had thirty-three failures before these three successes.

Bevis said he would publish his report in a peer-reviewed scientific journal. The head of the Royal College of Obstetricians and Gynecologists said it would conduct a full scientific study of how these births came about, and the ethics committee of the British Medical Association said it too would investigate and publish its findings. But Bevis's claim was never explored any further. No one ever found those three test tube babies, who had supposedly been born in early and mid-1973. And Bevis, on the recommendation of Lord Boyle, his university's vice chancellor, stopped working in the field altogether.

Doris Del-Zio's thirty-first birthday approached. Hank Aaron beat Babe Ruth's home-run record, nearly a quarter of a million people starved to death in the famines in Ethiopia, ballet dancer Mikhail Baryshnikov defected from the Soviet Union, the president of the United States resigned, and Doris could not get over the bitter disappointment of her failed IVF attempt. She gave up cleaning, marketing, cooking, doing the laundry. John prepared almost all the meals and did the grocery shopping, and Tammy did the rest. Sometimes John's or Doris's mother came in to help around the house.

Even her own friends and family had trouble understanding this passion Doris felt about wanting another child. She already had a daughter, whom she loved dearly, who gave her a reason to drag herself out of bed every afternoon and to do what she could to smile. Why, then, this yearning for an experience she had already had? That was a question asked of many women with what is known as "secondary infertility" — those who have borne a child but are unable to conceive another. So many of us, consciously or not, carry an image of a dream family, the right number or the right mix of babies that will at a certain point let us know that we have had enough. For people who are unable to complete their dream family, for whatever reason, the pain and disappointment, the sad sense of deprivation, can be as terrible as it would have been had they never had babies at all. And there are few social supports for secondary infertility. Observers are puzzled and disapproving, wondering what all the fuss is about and why the infertile couple can't just be satisfied with what they already have.

Finally, the Del-Zios decided they had suffered enough. They wanted to make someone else suffer, too. "It would have been one thing if it was a natural miscarriage," Doris said, "but this was a person who did this to me." In her long, sad history of miscarriages (including one during her first marriage that may have been brought on by her husband's physical abuse), a botched abortion (also during her first marriage, when her daughter Tammy was less than a year old), three failed attempts at artificial insemination with John's sperm, and all those unsuccessful operations to open her tubes, Doris had managed to accept her personal tragedies as acts of God. But this loss was an act of man. "I just can't accept that it was stopped by another man," she said, "a doctor who I had never known and who had done this against my baby . . . That was like deliberate and that is not fair."

At the urging of both William Sweeney and Landrum Shettles,

the Del-Zios sought the advice of a lawyer, Michael Dennis, the same lawyer who had represented Shettles in earlier disputes with Columbia. In the summer of 1974 the couple filed suit against Raymond Vande Wiele and his employers, Columbia University and Presbyterian Hospital, in federal court in lower Manhattan. Claiming that destruction of their potential test tube baby was "shocking, outrageous, repugnant," and "an intentional infliction of emotional distress," the Del-Zios sought a total of one and a half million dollars in damages — one million for her, half a million for him. It was a huge amount for that time, when the typical settlement in the case of a child's death through negligence was on the order of fifty thousand dollars.

At the same time that the Del-Zios were complaining that their IVF experiment should have been allowed to proceed, a group of leading American scientists was saying that their own experiments should be temporarily stopped. These scientists, who were engaged in a kind of molecular sleight of hand known as recombinant-DNA research, suspected that the knowledge they might generate, and the very methods of their investigations, were potentially so dangerous that they should be forced to stop for a while so they could catch their breath and think through the implications. The scientists' concerns resulted in a research moratorium — the same kind of moratorium that Congress had imposed on fetal tissue researchers looking for federal support, but with the crucial difference that the ban came not from Congress but from the scientific community itself. The only comparable action had taken place just three years earlier, when cardiac surgeons agreed to an international moratorium on heart transplants until they could figure out why so many of their patients were dying. But that was nothing compared to the scope of what was being proposed for recombinant DNA.

The kind of research the newspapers called "gene splicing" or "genetic engineering" involved taking bits of DNA from one species of microorganism and inserting it into the DNA chain of a different species, creating a new kind of organism, a crude patchwork of a thing. In 1974, when scientists first managed to insert a viral gene into a bacterium, such experiments suddenly started to look threatening — much more threatening, in fact, than IVF, which after all did nothing to alter the basic genetics of the species. With recombinant-DNA research, scientists became frightened by their own power to create new and potentially devastating hybrids for which there was no easy niche, no natural enemy, no obvious way to control.

They decided to stop what they were doing and take stock.

9
Science on Hold

> A good deal of genetic engineering looks to me as though one might be better off without it.
>
> — Herman Kahn, in *Can We Survive Our Future?* (1971)

The phone calls came in to Paul Berg's laboratory at Stanford almost every day, asking for pieces of DNA with which to conduct various gene-splicing experiments. Scientists were calling Berg because he was a leader in the emerging field of molecular biology, which involved cutting off a string of DNA at one particular point and inserting a string of DNA from a different microorganism in exactly the same place. The resulting "chimeras," as the scientists called them, were named in honor of the mythological beast made up of body parts of three different animals — the foreparts of a lion, the hindparts of a goat, and the tail of a serpent — cobbled onto a single monstrous creature.

Berg received the succession of phone calls with some alarm. "What do you want to do?" he would ask the scientist at the other end of the line, only to hear plans for "some kind of horror experiment." Then, according to Berg, "you'd ask the person whether in fact he'd *thought* about it and you found that he really hadn't thought about it at all." He started to worry about what was going on in gene-splicing laboratories across the country — beginning with his own laboratory in California.

At the urging of Berg and others, the National Academy of Sciences formed a committee to consider the consequences of genetic engineering. Eleven scientists — all of them men — joined; among them were one Nobel laureate (James Watson of Harvard) and four, including Berg himself, who would receive Nobels within the next ten years.

Scientists were already envisioning uses for recombinant-DNA technology. Not only would it be a big aid in basic scientific research to be able to insert new genes into familiar microorganisms and see what changed, but there would be dozens of useful applications in medicine and industry. Human genes that directed the production of essential proteins — insulin, growth hormone, clotting factor — could be inserted into bacteria, turning them into tiny drug-making machines. Genes inserted into bacteria that live in the mouth could prevent cavities from occurring. The nitrogen fixation genes that are present in legumes could be grafted onto cheaper crops such as corn and rice to make them nutritionally sufficient for millions of Third World children.

But the projections of where gene splicing might lead were troubling, too. "We must remember that we are creating here novel, self-propagating organisms," said Robert Sinsheimer of the California Institute of Technology. "And with that reminder, another darker side appears on this scene of brilliant scientific enterprise." Sinsheimer worried, for instance, about inserting viral DNA into bacteria for the sake of pure research, especially DNA of viruses that caused cancer in animals. If these lab-bred chimeras somehow got loose, he said, they carried an "almost completely unknown potential for biological havoc."

In late July 1974, the committee, with Paul Berg as chairman, wrote a letter to the editor of *Science* proposing a temporary moratorium on all recombinant-DNA experimentation. At that time in American history, the word "moratorium" usually brought to mind the efforts of college kids to bring ordinary commerce to a halt for a "Moratorium Day," a day devoted to teach-ins, marches, and demonstrations opposing the war in Vietnam. Berg and his colleagues suggested a time-out period, too, but in this case the moratorium on potentially dangerous work was called to give them time to assess just how dangerous it really was. Like the physicists of a generation before, who saw the terrible power of their science in its potential to explode into the atom bomb, these biologists were afraid to unleash the monsters that might lurk in their cell preparations before they knew what they might be getting into.

As soon as the Berg committee's letter was published, the gene-splicing moratorium — a completely voluntary pause monitored through a sort of honor code among molecular biologists — began.

The following month, a year almost to the day after Vande Wiele's confrontation with Shettles over the Del-Zio laboratory specimen, nearly two hundred representatives from thirty-eight countries convened in Paris on the campus of the Sorbonne to attend a conference called, grandiloquently, "The Future of Man." One of those from the United States, part of a national delegation of thirty-two scientists and scholars, was Raymond Vande Wiele.

When he took his turn at the roundtable discussion about human procreation, Vande Wiele seemed at first to have transmogrified into an IVF advocate. It would be fine to go ahead with in vitro fertilization, he said, as long as the sanctity of the family was preserved: the egg and sperm must come from a married couple, and the embryo must be implanted back into the wife's uterus. The situation Vande Wiele was describing as nonobjectionable was, significantly, just what had existed in the case of Doris and John Del-Zio. Did this mean he would now approve of such an attempt? Had he completely reversed his thinking about IVF in a single year?

As he talked, Vande Wiele added more and more caveats, leaving the distinct impression that he hadn't changed his mind after all. He offered one crucial condition that he thought had to be met. No IVF should be allowed, he said, until "it can be demonstrated that the procedure is safe; mainly that there is no increased incidence of fetal abnormalities." That was much the same condition he had offered during the Del-Zio debacle, a condition that was still nowhere near being met.

How could that level of safety ever be demonstrated to anyone's satisfaction? As Robert Edwards, who was also on the panel in Paris, said, "Any medical treatment is likely to cause a certain incidence of deformity. The unexpected can always happen."

Yes, Vande Wiele agreed, but he argued that certain safeguards could at least minimize the risk, such as requiring extensive testing on primates before allowing embryo transplantation in humans. Another safeguard would be the imposition of strict new protections for the embryo, since the most common protection, the research subject's fully informed consent, was obviously impossible when the subject was a day-old zygote.

"Unless we set up limiting regulations," Vande Wiele said pointedly, "there is the danger that unscrupulous individuals, because of their desire to be first, will take unacceptable risks." Did he mean that as a dig at Edwards, sitting on the stage with him at the Sorbonne? Or was he thinking perhaps of another "unscrupulous individual" in particular, of Landrum Shettles and that chocolate-colored mixture of gametes that had been warming, the previous September, in the incubator of a cramped, old-fashioned New York City lab?

Though the Paris conference was billed as "The Future of Man," a gathering that took place a few months later, on a wooded cliff on the coast of California, was the real historic event. Never before had there been anything quite like the Asilomar conference. For a few days in late February of 1975, 150 scientists assembled in the redwood chapel of a bucolic conference center to talk about how new life forms could be safely created and manipulated. "Debating the ethics of human interference with the mechanics of evolution in a church at the edge of the immense saline test tube where it all started: Rarely does one find one's metaphors so cheap, or so apt," wrote Michael Rogers, one of the few journalists covering the meeting, in *Rolling Stone* magazine. Asilomar was a remarkable moment, a happening, as stunning in its way as the natural spectacle with which it coincided — the annual return to this same forest of hundreds of thousands of monarch butterflies, a magnificent profusion of brilliant orange and black that every February turns the pines and redwoods of the Monterey Peninsula into living Tiffany lamps.

In the hypothetical experiment under discussion at the conference, DNA from a virus known as SV40 would be added to the nucleus of the bacterium *E. coli.* SV40, found in the kidneys of monkeys, is known to cause cancer in some mammals; *E. coli* is a usually harmless bacterium found in the human intestinal tract. The *E. coli* would be a specially attenuated laboratory strain known as K-12, which was unlikely to survive if it escaped from the lab and unlikely to be infectious even if it did escape. But the idea of mixing any kind of cancer-causing DNA into any kind of human microbe was sobering. The hypothetical SV40/*E. coli* experiment, which occupied the minds of the Asilomar participants for four days, was one that Paul Berg had planned to do before he helped draft the original moratorium letter for *Science.* "The Berg experiment scares the pants off a lot of people," one scientist observed, "including him."

The Asilomar conferees were focusing on a matter of great importance to molecular biologists: the host-vector pair. The vector in this sense of the term was a circular piece of bacterial DNA, known as a plasmid, into which a bit of foreign DNA was inserted. When a scientist introduced this new recombinant plasmid into an ordinary bacterium — most commonly the K-12 strain of *E. coli* — it would be taken up by the nucleus. If all went well, this souped-up *E. coli,* acting as the host in the recombinant-DNA system, would start expressing the foreign DNA by performing whatever task its genes dictated, such as producing a particular protein. The use of an enfeebled vector as well as an enfeebled host was thought to minimize the risks of an experiment like this — the kind of experiment that could some day lead to the microscopic pharmaceutical factories that people were envisioning at the time.

The scientists at Asilomar were gleeful about what recombinant DNA research could teach them and what tricks it could perform. Robert Sinsheimer of Caltech noted their "impetuous eagerness" to get back to their experiments, which he thought of as mostly a good and inspiring thing. Still, he worried that no one had left any time on the crowded agenda to discuss the most important issues. The conferees spent a great deal of time focusing on the benefits of recombinant-DNA research and on ways to assess its risks. They pondered whether gene-spliced organisms could escape from the lab and whether they could then survive and propagate; whether splicing genes of cancer-causing viruses presented particular risks; and how these potential hazards could be minimized or eliminated.

But no one at Asilomar was asking the larger philosophical question: should these experiments be conducted at all? That was a reasonable question for the public at large to wonder about, and Sinsheimer thought it belonged on the conference agenda. Did scientists have some inherent right to pursue their curiosity wherever it might lead? Did the right to free inquiry take precedence over the public's need to restrict it? Could recombinant-DNA research ultimately derail the course of human evolution — and, if it could, what would be the ramifications? Who could foresee the consequences of a manmade shift in the human gene pool? "And if we cannot foresee the consequences," Sinsheimer asked, "do we go ahead?"

It turned out there was time on the agenda to ask these bigger questions. After dinner on the third day the presentations at last went beyond the purely scientific to consider ethical, social, and legal issues. Alexander Capron, a young lawyer from the University of Pennsylvania, was asked to talk about how to impose regulations on gene splicing through NIH advisory committees and other mechanisms. He knew he had nothing reassuring to tell these scientists, so he began with a joke. A scientist and a lawyer, Capron said, were arguing about which was the older profession. Science is, said the scientist, offering Hippocrates and Maimonides. No, it's the law, said the lawyer, naming Pericles and Hammurabi. They looked for more and more ancient signposts to prove their points, at last reaching all the way back to creation.

God must have been a scientist, said the scientist, because he brought order out of chaos.

Well, maybe, said the lawyer. But where do you think the chaos came from?

The joke went over well — after all, the lawyer came off looking worse than the scientist — but Capron's presentation did not. He questioned the entire premise of self-regulation, the entire assumption behind the Asilomar conference. "This group is not competent to assign [the] overall risk" of gene splicing, he told the scientists. That task is the public's responsibility. And it doesn't matter that lay people don't understand the scientific complexities involved; they still should be entrusted with regulating scientific inquiry. "It is the right of the public to act through the legislature," he said, "and to make erroneous decisions."

Capron's warning was sobering, but it didn't change the trajectory of the meeting. Berg and his associates had decided that a written report would emerge from the conference, and they stayed up to the wee hours that third night compiling a five-page statement summarizing the consensus. The next morning, the fourth and last day of the conference, Berg announced that he wanted the statement discussed, amended, and accepted before noon. The goal was to state exactly how much risk was involved in recombinant-DNA research, both to the investigators and to the community at large, and to propose mechanisms for minimizing it.

One big concern was that any statement by the conferees might be enshrined — or, as Joshua Lederberg of Stanford put it, "crystallized" — in legislation, which could impose overly tight controls on a science that might turn out to be harmless. But Berg worried that Lederberg's warning would only paralyze the scientists and make it impossible to come up with a document at all. "If our recommendations look self-serving, we will run the risk of having standards imposed," he said, trying to get the participants to agree on *some* statement. "We can't say that one hundred fifty scientists spent four days at Asilomar and all of them agreed that there was a hazard — and they still couldn't come up with a single suggestion. That's telling the government to do it for us."

No one was completely satisfied with the consensus statement that emerged from Asilomar — but then, that's the nature of consensus. The conferees were not excessively focused on their points of disagreement, however; they mostly wanted to construct a document, before they adjourned, that would allow a national debate on recombinant-DNA research to begin. They never meant it to be the final word.

In large measure their statement did indeed touch off a national debate. It began on Capitol Hill in April 1975 at a hearing of the Senate Health Subcommittee, chaired by Senator Kennedy. Kennedy was a longtime booster of scientific inquiry; it was he who had protected fetal research by turning a research ban into a moratorium — different from the recombinant-DNA moratorium in that it was mandated by Congress, but similar in that it was meant to be temporary. But the gene-splicing discussion seemed to spook him. For the first time in his political career, he was asking questions that sounded hostile to the scientists appearing before him.

"Have there been any accidents, or have there been any problems in controlling the material?" he asked one of his witnesses, Stanley Cohen of Stanford.

"In controlling which material?" asked Cohen.

"The plasmids."

"I am not aware of any accidents," Cohen said, "or consequences of any."

"There has not been any diversion of the material at all?" Kennedy persisted, using terminology that made it plain how far apart the scientists and the politicians were, how hard it was even to find a common vocabulary.

"Any what, Senator?"

"Diversion," Kennedy said again. "Has anybody gained access to this material who should not have? Has there been any that has been misplaced or removed or taken that you know about?"

Cohen tried to explain how inappropriate a concept like "misplaced" was when talking about bacterial plasmids. "Plasmids are naturally occurring genetic elements that are in many bacteria found in nature," he said. "Since bacteria and plasmids are self-propagating, they can't be considered in terms that imply a specific quantity."

Although Kennedy soon dropped this line of questioning, he did not seem to understand the scientists' view of the risks of gene splicing. "Scientists alone decided to impose the moratorium," he said in a speech at Harvard in May, as though they had done something wrong, "and scientists alone decided to lift it."

One member of the committee that had written the moratorium letter to *Science* saw Kennedy's comment as proof that he "misunderstood the whole process." The moratorium, said David Baltimore of MIT, could be explained in just a few words: "We were the only ones who knew what was going on, and our whole point was to alert the public."

Now that the public had been alerted, the government was bound to devise a strategy to protect it from the possible risks of recombinant DNA — while allowing biologists to continue their work in the most exciting field of the day. In late 1975 an NIH committee comprising scientists, lawyers, and other scholars announced such a strategy, one that depended on two kinds of containment: biological and physical. Biological containment meant making the host-vector pair so weak that even if it escaped from the laboratory, it could never survive. Physical containment meant making sure no pair ever escaped in the first place. Biological containment was based on the strain of bacteria used as a host, abbreviated as EK-1, EK-2, or EK-3 (EK was short for *E. coli* K-12) in ascending order of risk. Physical containment involved requiring technicians to use gloves and masks and pressurized suits, installing barriers and negative air flow systems in the labs, and setting up a carefully defined chain of steps for disposing of potentially hazardous laboratory waste. The level of hazard ranged from P-1, which was an ordinary open lab, to P-4, a high-security, reverse-air-flow, spacesuit lab reserved for the most hazardous material. There were only a handful of such labs in operation in the United States at the time — at Fort Detrick, Maryland, where biological warfare research was first conducted, and at a few elite universities studying such deadly organisms as anthrax, smallpox, and Ebola.

If an experiment was very risky (for instance, because the viral DNA being transferred could cause cancer in humans if it escaped), it required a higher level of containment. For any particular experiment, containment could be achieved in a few different ways by juggling both the biological and the physical factors. If a scientist wanted to conduct a certain moderate-risk gene swap using an EK-3 host-vector pair, say, he might be able to do so in a P-2 lab. But he could get away with using a lower level of biological containment (EK-2) if he beefed up the level of physical containment (that is, a P-3 lab). The system was arcane, and any investigation involving recombinant DNA required approval from a national NIH committee, layer upon layer of approval and forced delay, before it could proceed. But at least the plan assured the public that potential hazards were being addressed — and at least it got the research moving again.

With recombinant DNA, scientists worried about the physical risks to themselves and to the community outside their labs. With IVF, on the other hand, scientists worried less about physical risk, such as the possibility of causing chromosomal damage in the babies, than about the ethical implications of creating life in the lab.

The National Commission for the Protection of Human Subjects of Biomedical and Behavioral Research, the one established by Senator Kennedy's bill, was focusing on those ethical implications while the Asilomar conference was taking place. Under the chairmanship of Harvard obstetrician/gynecologist Kenneth J. Ryan, the commission included physicians, lawyers, professors of philosophy, and professors of religion. One member, Albert Jonsen of the University of California at San Francisco, was specifically identified as a "bioethicist," but like John Fletcher, he had started out as a cleric; he was a Jesuit priest with a doctorate in religious studies from Yale. Jonsen taught in the departments of philosophy and theology — a typical route of entry into a field that had not yet found its footing. No one at the time, not even Jonsen, had formal qualifications as a bioethicist; the Georgetown program in bioethics had yet to graduate a single Ph.D. But over the years the same individuals were heard from again and again on presidential commissions, at congressional hearings, and in meetings across the country — Fletcher, Jonsen, Ryan, Kass, Ramsey — and they became, de facto, the first wave of experts, who would set the tone for the development of the field.

The commission's first order of business was to consider the ethics of fetal research and to write a report that would recommend either extending or lifting the fetal research ban that had been imposed by Congress in 1974. The commissioners met for two days, all day Friday and most of Saturday, every month for five months. They contracted with outside experts to prepare fourteen reports about the historical, ethical, legal, and scientific aspects of fetal research. They heard oral testimony from twenty-three witnesses, representing scientists (the American Society for Experimental Pathology, the Society for Pediatric Research, the Society for Gynecological Investigation), feminists (the Women's Legal Defense Fund, the National Abortion

Rights Action League), groups on the political left (the American Civil Liberties Union) and on the right (the Association of American Medical Colleges). Several witnesses came from antiabortion groups, such as the National Youth Pro-Life Coalition and the Maryland Action for Human Life. The man from the Coalition for Life urged the commissioners to make their decision about the ethics of fetal research according to the golden rule.

Monsignor James McHugh of the U.S. Catholic Conference presented as clear a description of the Church's position as had been heard in a long time. Since a human being exists from the moment of conception, he said, research on the fetus should be governed by the same ethical norms as those governing research on any other human subject. But one of the underpinnings of these ethical norms was informed consent. Obviously, the fetus couldn't give it, and to ask for proxy consent from the mother — the woman who had already decided to extinguish the fetus's life — would be, said McHugh, "a mockery." He said the only fetal research the church could support was research intended to benefit the fetus involved, and then only if it were conducted in a way that respected the fetus's "rights and dignity."

Right on deadline, in May 1975, the National Commission issued its report recommending that fetal research be allowed to proceed, with certain provisos. Even research on fetuses scheduled for abortion was acceptable, the commission said, if it would generate information that could not be obtained any other way; if animal work had already been done to answer similar questions; if the research carried little or no added risk to the fetus; and if the mother had given her informed consent, with the father not objecting. For research conducted on the fetus outside the womb, the commission added a few more restrictions: that the fetus be less than twenty weeks gestational age; that no changes be made in the abortion procedure for the sake of the research; and that "no intrusion into the fetus [be] made which alters the duration of life."

But even though the report found federal support for fetal research ethically acceptable, the ban remained in place. For one thing, the recommendations had to wend their way through the maze of the regulatory process: review by the secretary of HEW, publication in the Federal Register within sixty days, one hundred and eighty days for comments from the public, and then a prompt decision by the secretary — who was required, if he decided to reject the commission's recommendations, to explain himself in writing. But something else slowed the lifting of the moratorium, and this was a more intractable problem. One of the commission's suggested safeguards was that all research involving human fetuses had to undergo "ethical review" by a national Ethics Advisory Board. In 1975 no such board existed.

So fetal research was stalled. And what about IVF? It had never been made an explicit part of the fetal research ban, but the federal government was withholding grant support, just in case — and also requiring that IVF research undergo ethical review by the same nonexistent Ethics Advisory Board. The resulting research limbo trapped one embryologist, Pierre Soupart of Vanderbilt University, who had been trying to get an NIH grant to study IVF for more than two years. Soupart had asked for $375,000 to fertilize four hundred human eggs — more than had been fertilized by all the investigators in the world put together at that point — and to examine every one of them for chromosomal abnormalities. After subjecting the pre-embryos to DNA analysis, Soupart planned to destroy them. He had no intention of trying to turn them into babies. His experiment was designed simply to answer the question everybody was posing about IVF: Would manipulation of human gametes in the laboratory damage the genes?

Soupart believed it was premature to create human test tube babies until more was understood about the health and viability of lab-fertilized human zygotes. He was especially troubled by a report by M. C. Chang, then at the Worcester Foundation in Massachusetts, about a litter of rabbits created by in vitro fertilization in which half had abnormally small eyes. The prospect of manufacturing hordes of small-eyed human babies — with possible implications for the way those babies could see, the status of their brain connections via the optic nerve, and a whole raft of other worrisome uncertainties — was enough to frighten Soupart into writing up his grant application.

But the issue of small eyes turned out to be a red herring. Chang soon realized that the abnormality was occurring in half of his rabbits generation after generation — the perfect Mendelian ratio for an autosomal dominant trait. Small eyes, it turned out, were the result of a genetic variation in these rabbits, a trait carried on a single gene, and half of the rabbits in any litter were statistically likely to inherit that single gene from one or the other parent, no matter how they were conceived. By the time Chang figured that out, however, Soupart was already on the ethics review merry-go-round.

In July 1975, HEW Secretary Caspar Weinberger lifted the ban on federal funding for fetal research. Federal grants could even be used, he said, to support research on fetuses delivered after abortions as long as they were not kept alive artificially. Indeed, Weinberger said there might even be some instances in which the fetuses *could* be kept alive artificially, if the goal of the research was to find better ways of keeping premature babies alive.

Observers were stunned by Weinberger's position. Allowing research on aborted fetuses, even with restrictions, could be construed as a kind of political suicide, especially with the memory of that ghoulish experiment on decapitated fetuses still so fresh. But Weinberger protected himself from criticism by following most of the advisory commission's recommendations, which included requiring that all fetal research proposals be subjected not only to ethical review at the hospital or university level but also to the approval of a national Ethics Advisory Board. He recommended, too, that all research proposals involving human IVF be subjected to the same national ethical review.

Weinberger did not seem especially concerned that there was

still no national Ethics Advisory Board, and he left HEW the following month. So a moratorium that was meant to have been lifted turned into a *de facto* ban, as scientists were forced to wait for the required regulatory bodies to be put into place.

To much of the general public in the late 1970s, reproductive technology became conflated with gene splicing. It made a certain kind of sense to consider scientists who created new forms of chimeric life and those who created life itself to be two sides of the same beast, each one likely to brew something equally horrifying. But they were not the same. Scientists attempting laboratory fertilization of human eggs were never interested in trying to do what the molecular biologists were doing — mixing up the genes of different species to create organisms never before seen in nature. The public at large had trouble seeing the distinction, trouble realizing that even though IVF was artificial in the way eggs and sperm were handled, it was creating nothing new — just a human baby that was the genetic combination of one woman and one man.

There was an important political distinction as well. Unlike the molecular biologists, who had made a great and public show of their fears about the research they were doing, the IVF researchers were taking a deliberately low profile. They said little about the likelihood of chromosomal damage or developmental abnormalities that might result from their genetic manipulations — partly because they didn't know what the risks really were. So most people didn't pay much attention to what was going on in the IVF labs — even though, at least in Great Britain, scientists were closing in fast on the brass ring of the world's first test tube baby.

10

The First One

> Did I request thee, Maker, from my clay
> To mould me Man, did I solicit thee
> From darkness to promote me?
>
> — John Milton, *Paradise Lost* (1667)

Lesley Brown's story was not so different from Doris Del-Zio's — except for how it ended. By the time she made her way to Patrick Steptoe's consulting rooms, Lesley, a housewife and former "café girl" from Bristol, was twenty-seven years old; her husband, John, was thirty-two. Like the Del-Zios, the Browns were raising a daughter together, John's thirteen-year-old daughter, Sharon, who had been left in his care when his first wife ran off with another man. Lesley, a chubby woman with bobbed blond hair and delicate features, loved Sharon and had cared for her since she was a toddler. But she still wanted a baby of her own; her private dream of a perfect family was still incomplete. "I had never had a baby to look after," she said. "I longed for all those things that mothers usually moan about, like coping with dirty nappies and being kept awake at night."

After nine years of trying to get pregnant Lesley decided that something was wrong with her. The uncertainty over her fertility — especially in the face of John's apparent ability to sire a child — made Lesley build an emotional wall between herself and John, a big, curly-haired lorry driver who occasionally found himself tempted by other women. "Go on, go to her!" Lesley shouted when she discovered one of John's affairs. "Find yourself a normal woman."

By late 1976 the Browns had reconciled, and Lesley finally believed her husband when he told her they were in this fix together.

They arrived early at Steptoe's office on St. John's Street, chilled to the bone after getting lost on the walk from the Manchester train station on a sleeting November afternoon. Inside the consulting room, their situation looked more promising: Steptoe offered Lesley an "implant," his shorthand for in vitro fertilization. She jumped at the chance. "It just didn't occur to me that it would almost be a miracle if it worked with me," Lesley said. "I wouldn't have believed it if Mr. Steptoe had told me straight out that, after years of trying, no one had ever had a baby from an implant."

In those days it was generally left to scientists to choose what to tell — or not to tell — potential research subjects. Steptoe could be a distant man, efficient sometimes to the point of rudeness, and his coolness sometimes crackled in the heated context of an infertility consultation. Yet his sincere belief in the viability of this approach occasionally led him, cool demeanor or no, to an enthusiasm that could be positively disarming — and sometimes even a bit misleading. Whether to spare his patients anxiety, or to avoid scaring them off, or simply because he earnestly believed in his eventual success, Steptoe spun his version of the truth the way clinician-investigators often did and still do: he accentuated the positive. In this regard Patrick Steptoe, who was to become a mythic figure both to his grateful patients and to the world at large, might not have been so different from Landrum Shettles, whose reputation never fully recovered from his early thwarted attempt at IVF. Both men were driven to take the next step in their experimentation no matter what promises they might have to make along the way.

Joseph Califano, secretary of HEW under President Jimmy Carter, finally got around to commissioning an Ethics Advisory Board in 1977 — two years after his predecessor had issued the order to create one. Had Califano been stonewalling, looking for a roundabout way to make sure the federal government never paid for the IVF research that he found personally objectionable? A practicing Catholic, Califano recognized that the procedure might be a source of hope for infertile couples, but he said that personally he was troubled by the slippery slope. "Does the perfection of these techniques create a potential for abuse so severe that the federal government should not support or should strictly limit its support of the research?" he asked in his mandate to the Ethics Advisory Board. "Can techniques of in vitro fertilization and transplantation of the embryo damage the resulting fetus and lead to abnormal children? Will this research lead to selective breeding, to attempts to control the genetic makeup of offspring or to the use of 'surrogate parents,' where, for example, rich women might pay poor women to carry their children?"

Califano named a friend from law school — James Gaither, a prominent San Francisco attorney — to head the advisory board. Since Gaither had little expertise in reproductive technology or other aspects of the science that the board might be asked to consider, Califano named as vice chairman an experienced and widely respected biologist, David Hamburg, president of the Institute of Medicine, a division of the National Academy of Sciences. The rest of the thirteen-member board was heavily weighted toward medical expertise, with seven M.D.s (including Hamburg), one businessman, one philanthropist, one layperson (one of only two women on the board), one priest, and one philosopher. Gaither was the sole lawyer. Only two members had "ethicist" in their job descriptions: the priest, Richard McCormick of the Kennedy Institute, who was a professor of Christian ethics, and the philosopher, Sissela Bok of Harvard, who was a lecturer in medical ethics.

The board held eleven public hearings, one at NIH and one at each of HEW's regional offices, so members could hear what people across the country thought about IVF. What people thought, it turned out, was a little bit of everything. Some urged the board to encourage federal support of IVF research as a way to impose regulations on clinical research that was already under way without government oversight, using private funds. Some said IVF could never be ethical, no matter how much preliminary research was done on animals, because of the immorality of interfering with the natural process of human procreation. Some were troubled by the destruction of excess embryos that were not used for implantation; others said the problem of excess embryos was temporary and would disappear as soon as embryo freezing became feasible. To some witnesses, human life and personhood began at the moment of conception, meaning that the zygote and embryo, even in the laboratory, had the same moral status as a child. Some argued in favor of free scientific inquiry, no matter where it might lead; others emphasized how important it was to keep scientists' curiosity in check.

A few witnesses, according to the board's summary of the hearings, expressed concern "that the development of federal policy with such profound implications might be influenced by a public unnecessarily alarmed by inaccuracies and misinterpretations." It was up to the government, they said, to set the record straight and to provide fair and accurate information about scientific discoveries, no matter how controversial.

Some witnesses said that in an era of limited resources, there were better ways to spend the taxpayers' money — in particular, on more research into ways to prevent pelvic inflammatory disease and other conditions that led to infertility and created the demand for IVF in the first place. "There were also those," the board reported, "who believed it inappropriate to fund IVF when the majority [of Americans], they were certain, opposed such research on deeply-felt ethical or religious grounds. The government should not spend public money for experiments so clearly in conflict with the basic commitments of so many of the citizens."

The board also heard testimony specifically about the first proposal they were asked to review, the one from Pierre Soupart of Vanderbilt, who was still waiting for permission to examine the chromosomes of four hundred IVF embryos. The strongest testimony came from Soupart himself. "The essence of my proposal," he told the board, "is to establish the genetic risk involved in the obtainment of human preimplantation embryos by tissue culture methods — no less and no more."

Testing the safety of IVF on human gametes was essential, Soupart testified, no matter what the animal research indicated. "I do not believe that data obtained from laboratory animals, even from subhuman primates, can be extrapolated to human beings, especially when the extrapolation concerns chromosomes, which are very specific for different mammalian species." He said that if Edwards and Steptoe had waited for more preliminary work, the kind Soupart himself wanted to do, before forging ahead with embryo transfers, they could have avoided the great number of failures they had encountered — rumored to be anywhere from sixty to two hundred unsuccessful tries.

Soupart said he had no interest in attempting embryo transfer until he was convinced that human IVF did not carry unnecessary risk of chromosomal damage. "In my heart and conscience," he said, "as a scientist as well as a sensitive human being, I could not have conceived of a more ethical approach."

Leon Kass attended the board's first hearing, held in Washington, to ask them to look beyond the particulars of Soupart's proposal. He said that what was at stake with IVF was nothing less than "the humanness of our human life." Kass, whose self-education in ethics had already taken him to the National Academy of Sciences, the Hastings Center, and the University of Chicago, neatly laid out the various ways one could think about biomedical ethics. "For some people," he said, "ethical issues are immediately matters of right and wrong, of purity and sin, of good and evil. For others, the critical terms are benefits and harms, risks and promises, gains and costs. Some will focus on so-called rights of individuals or groups (e.g., a right to life or childbirth); still others will emphasize so-called goods for society and its members, such as the advancement of knowledge and the prevention and cure of disease." Kass advocated none of these approaches. What he wanted the board to do, he said, was to think through "the implications of doing or not doing . . . [and] to try to understand fully the meaning and significance of the proposed actions."

In Kass's interpretation of the slippery slope, the question was not whether a particular action was ethical in and of itself but whether it carried implications, and eventual applications, that were unethical. The immorality of a potential future application rendered the technique itself, here and now, equally immoral.

And what awaited us at the bottom of that slope? Witnesses testifying before the board saw many possibilities: that if we tolerated discarding abnormal IVF embryos, we would eventually tolerate ending the lives of other abnormal beings, such as newborns with severe birth defects or profoundly ill adults on life support; if we tolerated implanting only the healthiest-looking IVF embryos, we would eventually tolerate other forms of eugenics; if we tolerated creating human life in the laboratory, we would eventually tolerate creating entirely new species, bizarre hybrids and chimeras that were half animal, half man.

In the end the Ethics Advisory Board was unconcerned about the slippery slope. Some of the most dire predictions, they said — like human cloning or the creation of cross-species chimeras — were "of uncertain or remote risk." Others, like surrogate motherhood, were more likely, but they could "be contained by regulation or legislation." The board admitted that "there is an opportunity for abuse in the application of this technology as other technologies," but they concluded that, on balance, "a broad prohibition of research involving human in vitro fertilization is neither justified nor wise."

Howard Jones, too, was unconcerned about the slippery slope of IVF. He saw the procedure as so many of his scientific colleagues did: as a way to bring a happily-ever-after ending to a couple's story of infertility. But just as the field was getting exciting, it seemed, Jones was being forced to leave it. Johns Hopkins had a mandatory retirement age of sixty-five. The school had given Jones a two-year extension while he waited for his wife to turn sixty-five, but it was clear that both of them — after teaching at Hopkins for a combined total of eighty-five years — would be leaving their field of research.

At this fortuitous moment, in late 1977, Jones received a call from Mason Andrews, the creator of Eastern Virginia Medical School. Andrews was an old friend of Howard and Georgeanna Jones, having been their protégé during his internship and residency at Hopkins. Now he was calling his friends because his new school was in trouble. Applications were down, faculty members were scarce, and Andrews was the chairman and only member of the department of obstetrics and gynecology. The school had to get national attention if it was to survive, and he thought he knew how to do that. Rather than hire young, promising faculty members and hope for the best, Andrews took a lesson from the Boalt Hall School of Law in Berkeley, which had built a national reputation virtually from scratch by hiring retired law professors from the fanciest schools in the East. Officials of Eastern Virginia Medical School were about to set out on a similar path; Andrews's call to the Joneses was the first step. He asked if they'd like to move to Norfolk.

Move to Norfolk? Now? Just when they were about to close up shop and switch into retirement mode? What about all those things they had thought of doing in their golden years: writing, travel, archeological digs, sailing around the Chesapeake, and, most especially, spending time with their five grandchildren? The Joneses were unsure what to do, so they sought the advice of their three grown children, two of whom had gone into medicine themselves. Should we take on this challenge, they asked, or should we retire, as we had planned?

Do what you've planned, their sons and daughter said. You've worked hard your whole lives; you deserve a rest.

The prospect of watching scientific developments from the sidelines ultimately didn't sit well with the Joneses. Even the digs and sailing trips they thought they were looking forward to seemed somehow contrived to fill up time. In the end, over their children's objections, Howard and Georgeanna accepted the chance their old friend Mason was offering. It was a major upheaval: both of them had been born and raised in Baltimore, and they had never lived anywhere else, except for Howard's years as an undergraduate at Amherst College in Massachusetts. Georgeanna had attended Goucher, a women's college just outside Baltimore, and had gone on to medical school at Johns Hopkins, where she and Howard met (he graduated in 1935, she in 1936). They were Baltimoreans to the core — a breed that one of the city's favorite sons, the columnist H. L. Mencken, once described as "the yokel par excellence, the booby unmatchable, the king dupe of the cosmos." How would the Joneses manage in entirely new terrain?

Norfolk seemed to have little in common with Baltimore, other than the cities' historical tethering via the famous Baltimore-to-Norfolk clipper ships, such as *The Virginian.* Baltimore was a big, brassy, industrial hub, with barges and trains and factories enough to populate a little boy's dreams, while Norfolk was more low-key, its small harbor given over not to commerce but to the United States Navy. But there were enough similarities to make the Joneses feel comfortable about a relocation. Both cities were in the midst of a major effort at downtown revitalization. In the mid-1970s, Baltimore was making a great show of converting its seedy downtown harbor into a huge complex of shops, restaurants, hotels, walkways, and museums. And Norfolk was engaged in a similar project, demonstrating its throbbing civic heart.

Later, when asked why he had decided to reconfigure his life anew at the age of sixty-seven, Howard Jones always had a ready reply: "It was better than being put out to pasture."

Lesley Brown might not have heard the truth directly from Steptoe, but she did eventually piece it together on her own. That was not until months later, when she was already far along in her historic pregnancy. Her suspicions were raised when newspaper reporters started phoning her, started following John to work, started jostling one another outside the Browns' front door. With all this press attention, Lesley wondered how she could have missed the frenzied newspaper stories that must have appeared whenever Steptoe's other implant patients got pregnant. By the time she and her family were forced by the ravenous reporters to leave their home, she had finally figured it out. There had been no other articles because there had been no other pregnant patients.

"Why should newspaper reporters be chasing you?" Lesley's brother asked when the Browns knocked on his door looking for sanctuary. "You said hundreds of women had had babies like yours."

"I made a mistake," Lesley answered. "I'm the first one."

Seeing how avid the papers were for news of Lesley's pregnancy and the coming child, the Browns soon got into the spirit of the event. They sold exclusive rights to their story to a London tabloid, the *Daily Mail,* for a reported £300,000, the equivalent of more than half a million dollars. (John's first pay packet, as a fifteen-year-old apprentice at an iron foundry, had been just over two pounds threepence.) The deal gave *Daily Mail* reporters the right to shadow John day and night, but it meant that they would shield him from the rest of the paparazzi — their competitors. The *Mail* took its job seriously. The paper used nine different cars as decoys to ferry John around Oldham, hid him in the home of a photographer's mother in a fancy part of town, and had a flock of reporters living with him to record his every move. It even hired private security guards to stand outside Lesley's hospital room door.

The only thing that could have caused more media madness around an impending birth would have been if the Browns' baby had been created not by in vitro fertilization, but by cloning.

11

A Baby Clone

> We shall get at last a race of conditioners who really can cut out all posterity in whatever shape they please.
>
> — C. S. Lewis, *The Abolition of Man* (1943)

Cloning and in vitro fertilization are intimately linked in their technical aspects; cloning would not be possible without the laboratory techniques developed for IVF. The two procedures are linked, too, in the public imagination. That linkage has saddled IVF with a duffle bag of moral controversies that it might not have warranted on its own terms.

The difference between IVF and cloning is a crucial one, in terms of the extent to which the scientist is manipulating life. The difference starts with the very first step. Rather than mixing an egg and sperm in a test tube and hoping the egg will be fertilized, the cloner begins with a donated egg whose nucleus has been removed. Into that enucleated egg, the scientist inserts the nucleus from a body cell of the person to be cloned. The nucleus must come from a body cell rather than an egg or sperm cell to be sure that it has a full complement of genes, forty-six chromosomes' worth. Egg and sperm cells have only one-half the complement; the other half is acquired through fertilization. After inserting the new nucleus, the scientist shocks the egg in one of a number of possible ways — a literal electrical jolt, à la Dr. Frankenstein, is the technique most often used — to integrate the nucleus into the cell and to begin the process of cleavage. If all goes well, the egg develops into a normal embryo and a normal fetus, with one essential difference: unlike any other fetus, which is the product of the union of two half genomes, this one takes its entire genetic complement, intact and whole, from a single individual. Some people refer to cloning as "asexual reproduction" — the kind of reproduction we usually associate with amoebas and parameciums.

Cloning would be impossible without the refinement of IVF technology — just as those who warned of the slippery slope predicted. It was not so much that IVF itself would be problematic, but that learning how to do it would open the door to so many potential abuses. And the potential abuse most often invoked in those days was human cloning. "The test tube time-bomb is ticking away," wrote one London reporter, William Breckon, after Edwards and Steptoe's very first announcement of success in fertilizing eggs in the lab. It starts with IVF zygotes, he wrote; it ends, eventually, with the production of "a cohort of super-astronauts or dustmen, soldiers or senators, each with identical physical and mental characteristics most suited to do the job they have to do." Laboratory-bred embryos today, in Breckon's vision, meant throngs of tailor-made, narrow-minded human clones tomorrow.

The popular image of cloning came from novels like Aldous Huxley's *Brave New World,* in which eggs and sperm were united in hatcheries, split into numerous identical cultures, and grown under conditions that made some of them elite Alphas and others the hard-working lower classes, the Betas and Gammas — each group comprising masses of interchangeable beings. The image came, too, from movies like *Sleeper,* Woody Allen's futuristic comedy released in 1973, in which government officials tried to maintain social order by growing a replacement dictator identical to the original, right down to the last epaulette, by cloning a salvaged piece of the late dictator's nose. With Woody Allen as clown and Huxley as prophet, cloning was a terrifying prospect. All it took to convince someone of the horrors of the slippery slope was to say the word "clone."

Now, in 1978, on the eve of a lawsuit that would soon put IVF itself on trial, came reports that we were already partway down that slope. The news was that the worst possible application of IVF technology had already occurred — and that it was circuitously connected to the very man whose IVF efforts were about to come under scrutiny.

The story of the cloning was told by David M. Rorvik, Shettles's frequent coauthor and a science writer of national repute. In the book *In His Image: The Cloning of a Man,* published in the spring of 1978, Rorvik wrote about getting a phone call from a mysterious sixty-five-year-old millionaire, whom he called Max. A rich man in search of an heir, with no willing young wife available, Max thought that the only logical solution was to get himself cloned. And the man to do it, he believed, was Landrum Shettles.

Max must have done his homework. It was easy to see that Rorvik's and Shettles's fortunes had been tethered for years, ever since the two men wrote the popular *Your Baby's Sex.* In his many magazine articles about high-tech baby making in the 1970s, Rorvik heralded Shettles as "virtually the only researcher working on in vitro fertilization of human eggs," whose "daring experiments . . . have propelled mankind forward" by "contributing substantially to the study of human embryology." Rorvik made such claims about few other scientists, and few other journalists made such claims about Shettles.

According to the story told in *In His Image,* Max called only a few weeks after Shettles's contretemps with Raymond Vande Wiele. He had already heard about the trouble up at Columbia, Rorvik wrote, so there was treachery in Max's phone call — and in his specifically asking about Shettles. Could this be a devious ploy to set Shettles up for catastrophe?

Rorvik instinctively tried to protect his friend. He told Max that he had already discussed cloning with Shettles and knew that the doctor would not want to get involved. In truth, Rorvik wrote, he believed that Shettles *would* have been interested — and he lied about it to keep Shettles from making a calamitous misstep. "With the extreme polarization of views and allegiances that I knew had occurred in the wake of the [Del-Zio] incident," he wrote, "it was easy for me to imagine someone engaging in an attempt to discredit Shettles decisively. How better to achieve this than by involving him in a scheme to clone a multimillionaire — and then to reveal, moreover, that he was doing it all for money?"

Rorvik phoned Shettles almost as soon as he finished talking with Max — for technical advice only. Could human cloning even be done? he asked. Shettles was able to provide a comprehensive tutorial on the procedure. As he told a reporter five years later, when Rorvik's book was published, "After all, I've done everything but the cloning." He said he would have jumped at the chance to clone Max, would have done it for free if Rorvik had asked him, just because "I would get so much joy out of accomplishing it." But Rorvik never gave him the option — and anyway, Shettles said, his deepest desire would have been to clone not Max but Muhammad Ali.

Rorvik found a physician who agreed to clone Max — for a hefty fee, a cool million dollars — and in the book he used the evocative pseudonym Darwin for this doctor, after the creator of the theory of natural selection. According to *In His Image*, Darwin carried out his experiments on an island in the South Pacific, where such genetic manipulations were not against the law. He collected eggs from a series of native girls, all of them, according to Max's specifications, beautiful young virgins. Between visits to the island from Max and Rorvik, Darwin enucleated the eggs by spinning them in a centrifuge, put them into a nutritional broth, and added to each egg culture a nucleus taken from one of Max's liver or epithelial cells. Then he mixed in an irradiated (and therefore supposedly harmless) virus that was meant to stimulate fusion between the egg cell and the nucleus. Then he waited. One time — and one time only, despite dozens of attempts — the fusion worked, and one egg divided into a thirty-two-cell morula.

Darwin implanted the morula into the womb of an ethereal sixteen-year-old orphan virgin whom Rorvik called Sparrow. As Sparrow grew large with the baby that was genetically Max's identical twin, the millionaire fell in love with her, Rorvik wrote. His only quandary was whether he should marry her or adopt her.

As Rorvik told it, the baby was born in December 1976. Laboratory tests supposedly proved that the child was the genetic clone of Max, that the two were in effect identical twins grown in different uteri and separated by more than sixty-five years. The spooky implications of this accomplishment occupied newspaper accounts for quite some time.

Some observers didn't see what all the fuss was about. "Clone a human?" asked *Wall Street Journal* reporter Jerry Bishop. "Why bother?" To Michael Crichton, the best-selling author of *The Andromeda Strain,* who could clearly get carried away with a scary science story when he wanted to, cloning wasn't nearly as frightening as diseases arriving from outer space. Reflecting the naiveté of the time, he dismissed the notion of cloning as a problem so remote as to be uninteresting. "Superficially, cloning seems to provoke all sorts of moral and ethical quandaries," he wrote in a scathing critique of Rorvik's book in the *New York Times Book Review.* "But seen more carefully, it has inherently the ring of a trick, lacking both the intellectual importance of a basic scientific development, [and] the social importance of an important technological innovation. Ultimately, cloning is a masturbatory act."

Some people weighed in with armchair analyses of David Rorvik, who went into hiding as soon as the publicity began. "A fraud and a jackass," one scientist, Beatrice Mintz, called him in *Newsweek.* Added Leon Jaroff, his former editor at *Time:* "David is intelligent. David is a good writer. David is a little strange." And *New York Times* reporter Herbert Mitgang concluded that the book must have been fiction because Rorvik had made statements in the past "about his desire to write a novel called *The Clone*." Maybe, Mitgang suggested, *In His Image* was that novel.

Until that time Rorvik had been considered a respectable journalist. He had a master's degree from the Columbia School of Journalism and had been a science reporter for *Time* for two years. At thirty-four, he had been whipping out books at an amazing pace after the one he wrote with Shettles in 1970. He published two books in 1971, *Brave New Baby* and *As Man Becomes Machine;* one in 1973, *Decompression Babies;* and, in 1974, *The Woman's Medical Guide,* plus a novel, *The Sex Surrogates.* In 1976 he received an Alicia Patterson Foundation fellowship, one of the most competitive journalism fellowships in the country. And through it all he was writing articles for some of the country's top magazines: *The New York Times Magazine, New York, Esquire, Omni, Look.*

So why would Rorvik write a book that was so inflammatory, so likely to damage his hard-won reputation? Maybe he was grasping for the kind of fame that usually eludes serious midlist authors. If that was the case, it worked, emphatically. *In His Image* was debated on the front pages of the *New York Times* and the *Washington Post* for days after the first bound galleys started circulating. Even Johnny Carson made jokes about it on *The Tonight Show,* jokes like "After one scientist read the book he decided he wanted to try to clone Dolly Parton . . . [trademark pause] but he would have had to start out with two cells." And this one: "They've discovered that that book on cloning is not a hoax. In fact, the IRS discovered the doctor who did it. They discovered him when he tried to claim two thousand dependents." And "I can't believe the IQ of my neighbor's kid. Don't ask me how, but the other day he attached his Erector set to a TV Pong game and cloned ninety-seven copies of their cat. When his father asked him how he expected to feed them, the kid took a drop of milk and cloned a cow."

At the end of May, Congressman Paul Rogers of Florida, known as one of the staunchest supporters of scientific research on Capitol Hill, held hearings at which four leading biologists summarized the state of the art of cloning. They concluded that it could not be done in humans — not then, and perhaps not ever. Toads and salamanders and frogs had been cloned, they said, but no mammals — and the experiments in every species involved so many botched attempts that it would be highly unethical to try it in man.

One scientist, Peter Hoppe of the Jackson Laboratory in Bar Harbor, Maine, testified that he had done something akin to cloning: he had produced baby mice that had only one genetic parent. He and his associate, Karl Illmensee, had taken a mouse egg at the moment of fertilization, when the sperm nucleus and egg nucleus are still floating in two discrete packets known as pronuclei. They had removed one pronucleus from the fertilized egg and exposed it to an enzyme that stimulated the remaining pronucleus to duplicate itself, giving them two identical pronuclei. When the two identical pronuclei fused, as they do in normal fertilization, Hoppe and Illmensee had a mouse zygote with a full complement of genes, all of which were derived from a single parent. They then proceeded with all the steps of IVF: they grew the embryos in a petri dish, transplanted them into females, and produced seven healthy mice — five that had genes from a single female and two with genes from a single male.

Weren't those seven mice clones? Not really, Hoppe said. A clone would have precisely the same forty-six chromosomes as its donor, but these mice had two complements of randomly generated halves of twenty-three chromosomes — which is almost, but not exactly, the same thing. Hoppe said he wouldn't have true clones until he took the experiment one step further, repeating the procedure with the eggs of the females among these seven asexually produced mice.

Congressman Rogers asked Hoppe and the other three scientists at the hearing if they could state unequivocally that there would never be a mad scientist or a totalitarian dictator who would try cloning in humans. Every one of them said no, they could not deny that possibility. But that didn't mean Congress should act to choke off experiments now, they said. In a comment that would prove to be a foreshadowing of cloning controversies for the next thirty years, they said that any attempt to ban research through legislation "would only curtail future benefits and fail to prevent any misuses."

The furor over cloning sparked by *In His Image* built for a while, then died away. Rorvik came out of hiding and went on a tour to promote his book. But he refused to produce Max or the baby, now more than a year old, or even blood or skin samples that could be tested to see if their DNA was in fact identical. By summer more than one hundred thousand copies of the book were in print — a huge number for a nonfiction hardcover. Rorvik reportedly earned $390,000 from royalties and the sale of paperback rights.

In 1978 only someone living under a rock would have been unaware of the book about the baby clone. The same could be said about the test tube baby in England, whose birth was then being awaited with much anticipation and a good deal of fear — fear that it would be born deformed, fear that its life would be ruined, fear that it would signal the beginning of the end of the family as we knew it. And for Lesley Brown, the biggest fear was that something was going wrong with her revolutionary pregnancy.

12
Hang On

> Human destiny is bound to remain a gamble, because at some unpredictable time and in some unforeseeable manner nature will strike back.
>
> — René Dubos, *Mirage of Health* (1959)

The first sign of trouble came in mid-June 1978: Lesley Brown was not gaining enough weight. By then she was thirty-four weeks pregnant (of a normal term of forty weeks), but the baby's growth was lagging about four weeks behind. Not only was the baby small, its heartbeat was erratic. And one week later Lesley developed high blood pressure, which signaled the potentially dangerous condition called toxemia. Steptoe wanted to deliver right then — the only sure cure for toxemia is to end the pregnancy — but the baby did not seem big enough yet to survive.

When success had been just a theoretical possibility, Edwards and Steptoe had never considered that their lab manipulations might lead to any bizarre genetic problems. They had never really thought IVF would lead to chromosomal monstrosities; they had trusted that nature in her wisdom would spontaneously abort any grossly abnormal embryo. They knew that the vast majority of serious chromosomal abnormalities are "nonviable" — meaning incompatible with life — and they usually result in miscarriage in the first trimester, often before a woman even realizes she is pregnant. But in the summer of 1978, faced with a real live IVF pregnancy, with real live complications, Edwards and Steptoe began to fear the worst. Maybe, they thought, they had run into a dreadful bit of luck — the headlong intersection of their high-tech conception with a gestation that just happened to be going bad. Some 40 percent of all pregnancies, after all, encounter trouble at some point during the nine months. If two out of every five pregnancies have difficulties along the way, why should this one be exempt just because of its unique beginning?

Steptoe was more worried about Lesley's danger signs — her high blood pressure and low weight gain and her baby's small head — than he would have been in most patients; he knew how critical it was for everything to go absolutely perfectly. Any mishap, even something completely unrelated to IVF, could destroy the future of the technology altogether. Toxemia occurred in about 7 percent of all pregnancies, most often in women who were already suffering from diabetes, kidney disease, or high blood pressure. Steptoe could not imagine how the risk of toxemia could be affected by the way a fetus was conceived. But if he panicked and delivered the baby early to treat the toxemia, causing the world's first test tube baby to be a premie, with all the risks of abnormality associated with early birth — if all these catastrophes came tumbling down in response to Lesley's elevated blood pressure, that could spell the end of in vitro fertilization.

Two, three, even four times a day, the worried Steptoe called other senior gynecologists to consult about Lesley's latest blood pressure reading. "What if you were the Queen's gynecologist," one of them said, "with all the newspapers looking on? What would you do? You'd hang on, so long as there was no absolute danger signal that she was going to have a stroke or die. You must hang on."

So hang on he did — which meant that Lesley had to hang on, too. Steptoe admitted her to the hospital so he could monitor her more closely, registering her under the name Rita Ferguson in hopes of buying Lesley some privacy. He was concerned that she would be frightened by news reports of her condition and that the added stress of interviews would endanger her health. Indeed, when one tabloid newspaper did make its way into Lesley's room on a dinner tray, she burst into tears. TEST TUBE BABY ALMOST DIES read the headline. Lesley thought her baby had actually died and that no one had had the heart to tell her.

Reporters hovered around the hospital dressed as boilermakers, plumbers, window cleaners, and priests. A local detective agency bribed hospital workers for information. One enterprising person, said to be an American journalist, called in a bomb scare, necessitating the evacuation of the entire maternity wing — women with their newborns, women who had just had surgery, even women in labor — just for the sake, it would seem, of getting a look at this "Rita Ferguson," aka Mrs. Lesley Brown.

The media's appetite for the test tube baby story seemed to be infinite, bottomless, and cruel — and it radiated across the ocean, to the opening days of a trial in lower Manhattan that suddenly seemed to be the American analogue of the Lesley Brown story. The trial was being seen as an indication that if Yankee ingenuity had been given its head, America could have gotten there first.

Part Three

Test Tube Death Trial

13
Fooling Mother Nature

> All creation, including science, is a war against precedent. Science to be vital must grow out of competition between individual brains, foils one to the other, each man mad for his own idea.
>
> — Paul de Kruif, *The Sweeping Wind* (1962)

John Bower, the lead attorney representing Columbia University in *Del-Zio v. Vande Wiele et al.*, wasn't easily surprised. But he sure was surprised when he arrived at federal district court in Manhattan in July 1978 to begin jury selection for the Del-Zio trial. The size and vivacity of the crowd on the courthouse steps was like nothing he had anticipated. Three times before — once the previous November, two different times in May — he had mounted these steps prepared to pick a jury for this case. Three times the trial date had been postponed, and he had hurried back down the steps before lunch. On each of those earlier dates the press hadn't even bothered to show up.

No one had expected much of a turnout this time, either. Docket organizers had scheduled the trial for Courtroom 312, a smaller-than-average room tucked away on the third floor. It never occurred to them that the Del-Zio case might require a venue large enough to accommodate all the newspaper reporters, courtroom artists, and curiosity seekers who would want to get inside.

But now thirty or more reporters jostled Bower as he tried to make his way to the courthouse door. Two blocks from City Hall, one subway stop from Wall Street, a short walk from where the double exclamation marks of the World Trade Center had opened just five years before, the federal courthouse had the white marble look that Americans have come to associate with reasoned jurisprudence. The brightly clothed journalists balancing video equipment and huge microphones, shouting "Are you with the Del-Zio case?" hissing at the back of Bower's head as he issued a steady string of "No comment"s, added a cartoonish splash to the otherwise sedate scene.

Why was everything so different this morning? Because on this day — Monday, July 17 — reports were coming in that the world's first test tube baby, the same kind of baby that the Del-Zio trial was all about, was soon to be born in England.

Those reports changed everything. Suddenly "test tube baby" was a familiar phrase, and even those who thought the world's first might turn out to be a monster found themselves swept up in the drama, the uncertainty, the sheer spectacle of it all.

In 1978 a popular television commercial featured a formidable blond woman in a long white gown with a crown of flowers in her hair. Looking a bit like Demeter, the goddess of fertility, she was standing in a cornfield eating a muffin spread with what she believed to be butter. When she found out it wasn't butter but margarine — a loathsome product of manmade chemistry — her eyes flashed, the earth shook, the thunder roared, and we heard her bellow, "It's not nice to fool Mother Nature!"

Yet fooling, or at least fooling around with, Mother Nature was exactly what was happening in biological labs in the 1970s. Scientists were learning more and more tricks that allowed them to toy with the godlike role of bending the world to their will. And in 1978, the year of that margarine commercial, some scientific developments were coming to a head in ways that would make the year, in retrospect, look like a watershed. The world seemed to be a Pandora's box, and scientists seemed to have the curiosity, the tools, and the hubris to insist on opening it and peering inside, no matter what horrors might be unleashed in the process.

A decade and a half after the New York World's Fair, the four "ages" touted in the fair's vision of Tomorrowland had all been subverted. The Information Age was making ordinary Americans worry about issues of privacy and security; the Space Age was flirting with catastrophe, most dramatically with the close call of *Apollo 13;* the Consumer Age was delivering truckloads of poorly made, easily broken junk; and the Atomic Age was breeding dangerous near-misses, like the spontaneous combustion of plutonium rods at the Rocky Flats nuclear weapons plant. As for the "Wonderful World of Chemistry" theme of one of the World's Fair pavilions, chemistry no longer seemed to have much to offer in the way of wonder. It was known now as the science that brought us deadly pesticides and toxic wastes rather than the miracle of Teflon.

Did anyone still believe that science was the key to a golden, prosperous, trouble-free future? Well, maybe. The news of Lesley Brown's pregnancy made even the most cynical technophobe willing to believe, at least temporarily, in happily ever after. For a little while in July 1978, the whole world waited to see what kind of baby had been made, outside of its mother's body, over in Great Britain.

If the British baby really had been created through IVF — there was a certain skepticism about that claim, similar to the raised eyebrows that had greeted Douglas Bevis's report of IVF successes four years earlier — and if the baby really was healthy, which was not at all assured, then the proceedings about to begin in the Del-Zio trial would be affected. It would be easy for the plaintiffs to make it seem, in retrospect, that Shettles had been a prophet and Vande Wiele a scolding, censorious Cassandra.

What good timing for the Del-Zios, to have their trial join in the global parade that summer. The timing was so fortuitous that some people — specifically, the defense attorneys — thought it must have somehow been deliberately contrived. The couple first filed their $1.5 million lawsuit in 1974, against not only Raymond Vande Wiele but also against his two employers, Columbia University and Presbyterian Hospital. The fact that there were three separately named defendants, represented by three separate law firms, ensnared trial preparations. Depositions, interrogatories, pretrial motions — everything took three times as long because it had to go through three teams of attorneys. Even the scheduling of jury selection was more complicated because of the speaking engagements and travel commitments of all those busy people. The original date for the trial to start was November 10, 1977; then it was switched to May 2, 1978, then to May 11, and then — because Raymond Vande Wiele was scheduled to be abroad until the end of June — to Monday, July 17.

Over the weekend of July 15 and 16, news reports about the test tube baby birth watch in England led the defense attorneys to press for yet another trial delay — until a time when all the clatter over test tube babies had a chance to quiet down. "There have been extensive developments from Britain — perhaps coincidentally, perhaps not," said Columbia's attorney, Bower, in the judge's chambers that first morning, before court was in session. Bower was a bearish man with the unfortunate accent of Colonel Klink on the popular sixties TV show *Hogan's Heroes.* He had acquired the accent from a childhood spent abroad, but Bower was thoroughly American; he was even a war veteran, having served as a paratrooper in World War II. Bower was a brusque and offputting presence in the courtroom, tending to bore in on witnesses in a way that edged close to badgering, objecting to his opponents' lines of questioning on the slightest whim, even stamping his feet.

Isn't the timing of this trial date an odd coincidence? Bower asked Judge Charles E. Stewart. With excitement building in England, the newspaper headlines and TV shows were all screaming about the "test tube death trial" in downtown Manhattan. Didn't the news from abroad poison the chance of finding a dispassionate jury in New York? Bower tried to make the case that his opponent, Michael Dennis, the attorney for John and Doris Del-Zio, had somehow concocted the coincidence. Dennis had had a long personal and professional association with Landrum Shettles, and it was Shettles who had suggested that the Del-Zios hire him for the case. Now Bower was suggesting that Dennis, through his Shettles connection, might have somehow persuaded the British scientists to announce the impending birth ahead of schedule — on a date that would, conveniently, synchronize with jury selection.

At this veiled accusation Dennis, an experienced litigator with a famously short fuse, exploded. Dennis had as military a bearing as Bower — as an army colonel he had been involved in military intelligence during what he always called "the Second War" — but he had a more engaging, homespun American manner. He looked a bit like George Burns minus the cigar. But right now there was nothing especially engaging about him. He began by angrily reminding the judge that the trial had already been delayed on three separate occasions, beginning the previous November, and that each time it had been Bower, not Dennis, who had requested the delay.

"Are we to be denied our day in court as promptly as we can get it now," he asked, "simply because science is progressing in the same field and coming out with a dramatic announcement?"

Judge Stewart decided that jury selection should begin as planned. As for the odd linking of the trial date with the impending birth of a test tube baby, one thing was sure: whether coincidental or contrived, it was clearly a good thing for Michael Dennis and his clients. In July of 1978, it seemed that everybody was rooting for in vitro fertilization to work.

On the first day of the trial the attorneys empaneled a jury of four women and two men, including a business executive, a nurse's aide, and a reporter for the *New York Times.* Doris Del-Zio was sworn in as a witness, and for a long time she performed well on the stand, as though she had been rigorously coached. The three defense lawyers tried a variety of twists and digressions to trip her up, but Doris essentially stuck to her story, even when the story seemed a little bit strained.

There was, for example, the "trial run" of IVF that Doris said had taken place fifteen months before the real procedure. According to her testimony, William Sweeney at New York Hospital and Landrum Shettles at Presbyterian had walked everyone through the procedure in June 1972, using her own and John's actual gametes. Everything had gone smoothly, she said — leading the doctors and the Del-Zios alike to believe that the IVF they embarked upon in September 1973 would go smoothly too.

The trial run took place, Doris testified, when she was at New York Hospital for one of many operations on her damaged tubes. During the surgery, she said, Sweeney removed a wedge-shaped piece of her ovary, placed it in a test tube, filled another test tube with a sperm sample from John, wrapped the test tube carefully, and handed the package to his stepson, who was waiting in the atrium just outside the operating room. According to her testimony, the young man headed over to Presbyterian Hospital — he took a taxi, just as John Del-Zio would the following year — and approached a balding, sharp-featured man wearing a white coat, who was standing in front of the hospital entrance. "Are you Dr. Shettles?" Sweeney's stepson asked. "Yes," said Shettles — and, as the plaintiffs described it, the young man handed over the tightly corked test tubes and headed back across town.

When Shettles received the test tubes, according to Doris, he took them into a borrowed lab, mixed the Del-Zios' gametes together in a fresh test tube, and placed them in an incubator. A day later, she testified, the egg had been fertilized and had already split into two distinct cells. That pace was much faster than was typical; Robert Edwards had already established that generally the first cleavage did not occur until thirty-eight hours after fertilization. After the fertilized egg split, said Doris, Shettles and Sweeney were confident that an actual IVF attempt could work. According to their prior agreement, Shettles supposedly destroyed the trial-run embryo.

The thing about the June 1972 trial run was this: the defense lawyers thought it was a complete fabrication. And if they could convince the jury that it had never happened, they could establish two points: that Sweeney and Shettles were less sure about what they were doing in September 1973 than they pretended, and that Doris was lying.

The defense lawyers believed the plaintiffs had cooked up the trial-run story sometime between 1976, when Michael Dennis took Doris's deposition, and 1978, when court proceedings got under way. In her deposition Doris had said that the first time she had heard of in vitro fertilization was during a chat with Sweeney in his office in August 1972. How could August be the first time she had ever heard of IVF, asked Stephen O'Leary, who represented Vande Wiele, if she had supposedly gone through a trial run of that very procedure in June — a full two months earlier?

According to Doris it was a simple mistake. She might have said August in the deposition, she testified, but she had meant to say April. She said she often got April and August confused, just as she confused June and July, and all those other months that start with the same letters.

"After you read over the testimony in 1976," asked O'Leary, "did you call Mr. Dennis and say 'Mr. Dennis, there is an important error in this testimony'? Did you do that?"

"No, sir."

"And you knew when you read over this testimony there wasn't one word about this so-called trial run in June?"

"Sir, I wasn't asked if there was a trial run in June."

"But you knew about the trial run in June when you read it, didn't you?"

"They certainly never asked me if there was a trial run in June."

O'Leary began again: "You knew this trial run was in June; right?"

"Yes," said Doris.

"You knew that, right?"

"Yes," she said again.

"And when you saw the testimony [that] the first time that you knew about in vitro fertilization was in August 1972," O'Leary asked, "didn't that bring to your mind that there was an error in your testimony?"

"No, sir. In the deposition there was never a mention of a trial run."

On July 19, Doris's third day on the witness stand, O'Leary returned to the issue of the trial run. Probing the August/April confusion, he began, "Mrs. Del-Zio, would it be fair to say that since you were having this problem with not having a baby, that when you first learned of in vitro fertilization, this was a very important, significant day for you; isn't that right?"

"I said yesterday, sir, I was very happy when I learned of it, yes," Doris replied.

"And that was one of the most important, significant medical days throughout your medical history, isn't that true?"

"No," Doris said, refusing to be tricked. "My most significant day is not April, it is September 12, 1973." That was the date when Shettles was supposed to create an embryo that would eventually be implanted into her womb.

"All right," the lawyer said. "But wouldn't you say, Mrs. Del-Zio, knowing the past medical history that when you first learned of this in vitro procedure, that was one of the most significant days in this medical history?"

"No, sir, I can't say one of the most significant."

"One of the more significant, would that be okay?"

"I would say it would be an important day."

"One of the more important days?"

This kind of semantic bartering went on throughout the trial. "No, I didn't say 'more,'" Doris replied carefully. "I said it was an important day."

"Just an important day," O'Leary conceded. "We will adopt that word. An important day. And would you say, Mrs. Del-Zio, that the day you first learned about in vitro fertilization, that that would be a more important day than, let's say, the day you were discharged in 1970 in your hospitalization? Would you say that?"

"Yes," she answered, warily.

"Comparing those two days, isn't that true?"

"Yes."

"In your examination before trial, Mrs. Del-Zio, isn't it a fact that you knew the exact date you were discharged in 1970 from the hospital?"

O'Leary might have had her there. But Doris would not be had. "It was a lucky guess if I said it, sir," she said.

Of course, if the trial run did actually happen, it meant that Shettles himself had thrown away a successfully fertilized Del-Zio zygote — the very thing the Del-Zios were suing Vande Wiele for having done. In that case, the defense attorneys asked, hadn't Shettles contributed to the same anguish and despair that Doris said she was suffering at Vande Wiele's hands? If she kept insisting that Vande Wiele "killed my baby" in September 1973, couldn't the same be said of Shettles in June 1972?

"Those two specimens were both in the same biologic situation, were they not?" asked James Furey, the lawyer representing Presbyterian Hospital, during cross-examination.

"Not to me," said Doris.

"Not to you?"

"No, sir."

"Did you ever inquire as to what happened to the first specimen?" Furey asked, taking a slightly different tack.

"No," said Doris. "Dr. Shettles was taking care of that. You would have to ask him."

"Can you assume that Dr. Shettles killed the first specimen?"

"He didn't kill anything. He wouldn't kill anything."

"Well, did he dispose of the first specimen?"

"As I said, you will have to ask Dr. Shettles," Doris said. "I don't know."

Furey acted surprised. "You were not concerned to inquire about it?"

"I knew I did not have the implantation since my body was not ready for it," said Doris. To her mind, the specimen from the trial run was a laboratory artifact, and the specimen from the real procedure was a baby.

❧

The three defense lawyers often repeated themselves and one another, which occasionally unsettled the witnesses. Even Furey, the mildest of the attorneys, sometimes went too far. On Doris's last day of cross-examination, he hammered away repeatedly about the newspaper and television interviews she had given in the weeks leading up to the trial. Were there three networks? More? Did she cry on tape? Was her attorney present? The questioning grew so insistent that Doris finally uttered the not-quite-accurate but still affecting plea that made the next day's lead headline in the *New York Daily News:* "Please don't brainwash me anymore!"

But Furey's questioning was no match for Bower's. With his Germanic bearing and accent, Bower took on the role of bad guy almost from the opening day. On one occasion, while cross-examining Doris, he grew positively abusive. In reviewing her fertility history, he asked her to agree that she was essentially incapable of carrying a pregnancy to term.

"No, sir," she countered; "I have a fifteen-year-old daughter."

Bower began again. "Ma'am, since that time, since Tammy was born and a lot of water has gone . . ." He paused, apparently thinking better of implying, in front of a jury comprising mostly middle-aged women, that Doris, at thirty-four, was past her prime. "Well, a lot of changes have gone on in your life — right?"

"Yes, sir."

"As a matter of fact," said Bower, his voice now a booming crescendo, "you had a miscarriage when your first husband kicked the hell out of you."

The courtroom erupted, reporters scribbled leads for tomorrow's stories, and Michael Dennis was on his feet. "Objection, Your Honor," he shouted. "I think counsel ought to be admonished."

"All right, I withdraw that."

Dennis would not be mollified: "That is a disgrace."

Judge Stewart agreed. "Mr. Bower," he said, "that was entirely unnecessary."

"I will withdraw it."

"Your Honor," Dennis persisted, "I ask that this counsel, who should know better, ought to be admonished."

"I am sorry," said Bower, just as the judge, turning from one attorney to the other, said, "Mr. Dennis, please. Yes, Mr. Bower, I am really distressed at you."

"I am sorry, I will rephrase it."

"No, you will not do that," Judge Stewart insisted. "We will take a short recess."

The damage, of course, had already been done. The next day's *Daily News* carried the headline TEST-TUBE MOM ADMITS: LOST BABY WHEN BEATEN BY EX. Beneath a drawing of Doris on the witness stand — scarf tied around her neck, a bright white handkerchief in her upraised right hand — the caption took up the rest of the front page:

> Doris Del-Zio weeps, as she testifies during yesterday's $1.5 million test-tube damage suit in Manhattan Federal Court. Del-Zio, seeking damages against doctor after test-tube baby was lost, admitted yesterday that she had suffered a miscarriage 15 years ago after "a physical beating by my first husband." Defense lawyers

suggest that inability to bear children may have stemmed from earlier injuries. Stories on page 3.

Again and again during the trial, Judge Stewart reminded the attorneys to steer clear of the press. The jurors were not sequestered — this was, after all, only a civil case — so any quotes that appeared in the newspapers or on local news broadcasts were easily available to them. Every morning the *Daily News* headline blared its tabloid interpretation of events at the "test tube death trial"; every evening local TV news show carried dispatches from lower Manhattan. On the fifth day of the trial the jury foreman, Hugh Lawless, was spotted in Grand Central Station with the *Daily News* folded under his arm; he had clearly been reading it on his commute in from Dobbs Ferry. (The headline that morning read, TEST-TUBE MA'S SPIRIT WAS DESTROYED ALONG WITH THE EMBRYO.) Judge Stewart was the one who had seen Lawless in the train station, but he decided not to make a big deal about it. "I don't see any reason why he shouldn't be carrying a copy of the *Daily News*," said the judge. "He's as interested in how the Yankees did yesterday as I am."

On the witness stand Doris was strong and a little bit clever; she would not be tricked into saying the wrong thing. But to hear her tell it, she was a hapless victim, a shell of her former self, rendered too depressed and apathetic to look out for herself, her husband, her daughter, or her home. A simple woman with a simple desire to have a child of her own, she was besieged by nightmares and hallucinations about the baby she never had.

Doris's uncanny calm under cross-examination was slightly unsettling, given that her main complaint was that Vande Wiele's action had left her emotionally fragile and overwrought. But when she did occasionally break down and weep, some observers found themselves wondering which Doris was the real thing: the steely-eyed witness or the puddle of gloom?

Furey tried to highlight Doris's mercurial nature as a reason to question her sincerity. Why was she submitting herself to all this, he kept asking her, if the episode was as devastating to relive as she claimed? For just one reason, she replied: to keep other women from having to endure what she had gone through.

"In other words," Furey said, "you wanted to make sure that nobody would go through the heartache of this rash experimentation, is that right?" He spoke quickly, as if to get the phrase "rash experimentation" uttered in front of the jury before anyone objected.

But Doris had an image of her own she wanted the jury to carry away. "Well," she said, "I wanted to make sure that nobody had a Dr. Vande Wiele to kill their baby." *Kill their baby* — three words contrived to send a shudder through the courtroom.

"You know that wasn't a baby, don't you?" asked Furey, trying to minimize the impact of her words.

"I don't know that," Doris shot back. "That was my baby."

Doris was the main attraction for most of the tabloid press, but her gynecologist, William J. Sweeney III, was almost as colorful a witness. Doris used to call him a leprechaun: he was a short man, shorter than she was, with wavy hair, a distinctive cleft chin, an Irish lilt to his voice, and a tendency to crack bad jokes. The item of Sweeney's wardrobe most often remarked upon was his yarmulke. Sweeney was raised as a Catholic, but he was lapsed with a capital L — so lapsed that he had married three times, so flexible that each time he married he converted to his new wife's religion. In 1978 his third, and last, wife was an Orthodox Jew, and ruddy-cheeked, snub-nosed Sweeney, with a face that would be more at home in a neighborhood pub than a synagogue, wore a skullcap every day to please her. He kept kosher, too.

Sweeney was the kind of doctor who ate his lunch at the bedside of his hospitalized patients. He was well known in the ob-gyn community of New York and had once been a close associate of Raymond Vande Wiele. In fact, when Sweeney applied to be chief of obstetrics at St. Luke's Hospital, an affiliate of Columbia University, he asked

Vande Wiele for a recommendation. But Sweeney did not get the job, nor did he get a similar job he applied for at Roosevelt Hospital, also a Columbia affiliate. And after September 1973 the friendship deteriorated altogether. Sweeney phoned Vande Wiele about a year after their IVF confrontation to ask, "Are you still my friend?" Vande Wiele responded with a curt and hurtful "No."

Shortly before the Del-Zios' IVF attempt, Sweeney, then fifty-one, had collaborated with one of his patients, Barbara Lang Stern, on the book *Woman's Doctor: A Year in the Life of an Obstetrician-Gynecologist.* It was a popular book, with a breezy, conversational, relaxed style — perhaps a bit too relaxed. At the trial, defense attorney John Bower seemed to get a kick out of reading graphic passages aloud while Sweeney was on the witness stand. "What the hell," Sweeney had written disparagingly about natural childbirth, "if it is going to be really natural the lady ought to deliver a baby and eat the placenta." And "I don't generally take phone calls during office hours. Usually I can't, because I've got my hand in somebody's vagina." As Bower got ready to read a passage about Sweeney's patients' infidelities, Judge Stewart finally told him that enough was enough. Perhaps the judge had read the book himself and knew how many embarrassing sections there were to choose from if the reading went on. Noting that vaginal discharge and itch were the most common problems he treated, the Cornell professor of gynecology wrote, "It's easy to laugh or say, 'Oh my God, here comes another itchy twat!'" Clearly this was not the impression the plaintiffs wanted to leave, of Sweeney as a crude and vulgar man who referred to his patients' genitals as twats.

In the IVF collaboration with Shettles, Sweeney's scientific opinions took a back seat to Shettles's. But at the trial the lawyers wanted to hear what Sweeney thought about the risks and potentials of IVF, about the likelihood of creating babies with either obvious or latent abnormalities. They wanted to know, too, why Sweeney had never presented his plan to the Human Experimentation Committee at New York Hospital. That committee, like the one at Columbia, was part of the institutional review process to ensure that certain steps were followed to obtain fully informed consent from research subjects and to protect them from unnecessary harm. "I did not consider this an experiment," Sweeney said to explain his failure to alert the committee. "I considered it a surgical procedure."

As for its riskiness, Sweeney did not think IVF increased the likelihood of genetic damage; the chance of creating a chromosomal abnormality, he said, was "minimal or nonexistent." In fact, he testified, the genetic risk might actually be lower with IVF than with natural conception, "because we could examine [the embryo] microscopically before putting it into the uterus" and not transplant those with visible irregularities. In addition, he said, IVF bypassed the "extremely tortuous passage" through the fallopian tube that an early embryo usually endures, in which it is "squeezed and forced" in a way that could lead to problems. This was an interesting way of looking at the risks and benefits of IVF, but it was a perspective that few other scientists shared with William Sweeney.

Sweeney and Shettles and Doris Del-Zio — most especially Doris — believed that the test tube containing the Del-Zio gametes, the specimen that resembled a chocolate milkshake, was more than a froth of dividing cells, that it was in truth a woman's own baby. This fierce conviction, which Vande Wiele and the other defendants tried to dismiss as excessively romantic, was made to seem not so absurd, in fact even reasonable, just one week after the trial began — and all because of the revolutionary birth that took place late on a Tuesday night in England.

14
Pandora's Baby

> The first reaction of most people to the arrival of these asexually produced children, I suspect, would be one of despair.
>
> — James D. Watson, "Moving Toward Clonal Man" (1971)

On the afternoon of July 25, 1978, Patrick Steptoe decided it was time for the "Baby of the Century" to be born. Reporters from around the world converged on the rolling countryside of Lancashire, waiting for early reports of the historic birth. They camped out on the front lawn of John and Lesley Brown's house in Bristol; tried to sidestep the guards outside Lesley's door in the maternity wing of the hospital in Oldham; or, if they worked for the *Daily Mail*, accompanied John to work, to the hospital, and to the pub — access permitted them because their editors had paid the Browns a small fortune for exclusive rights to their story.

Lesley Brown was just over thirty-nine weeks pregnant; she had managed to hold on for almost the full forty weeks. After five weeks of bed rest, her baby's growth had nearly caught up to its gestational age. When a test of the amniotic fluid showed that the baby's lungs were mature enough for a successful transition to extrauterine life, Steptoe made plans to do a cesarean section that night.

But he didn't want the hordes of journalists to sully the event that he and Robert Edwards had been working toward for so long — an event that they, along with the parents, wanted to savor in privacy and peace.

Peace and privacy, of course, were by this point relative terms. Edwards and Steptoe had already granted permission to the Central

Office of Information, a division of the national Ministry of Health, to have a camera crew present to film the delivery. Figuring out the conditions under which the crew could be there and ownership of the film once it was made were now the subject of much last-minute negotiation.

For Steptoe, alone in Oldham while Edwards took a rare holiday in a remote Yorkshire cottage, the first hurdle was that Lesley needed to fast for eight hours to empty her stomach in preparation for surgery, and it was already nearly suppertime. How could they take Lesley off all foods without raising the suspicions of the hospital staff? Steptoe worried that if more than a handful of people knew about the impending birth, someone would leak it to the press, and chaos would prevail. So he enlisted the help of only two people: Edith Marshall, a nurse-midwife who had become a special friend of Lesley's, and his own wife, Sheena. Sheena carried a plastic garbage bag into Lesley's room and dumped her entire uneaten supper into it, then tied the bag up and hid it in another bag.

"You *have* eaten well," the ward sister said when she returned for the empty tray. "Are you still hungry?"

"No, thank you," said Lesley, stifling a conspiratorial smile. "I've had quite enough."

Then, what to do about John? If he started behaving differently as his wife prepared for her cesarean section, the *Daily Mail* reporters at his elbow would surely figure out that something was up. Steptoe decided for the moment to tell John nothing, but to let him visit Lesley in the evening and leave at his usual time, about nine. The operation wouldn't begin until eleven at the earliest, given the hour of Lesley's last meal, so there would be plenty of time to summon John back to the hospital. This way his entourage of reporters might be put off the scent.

Edwards was not expected back from Yorkshire until eight that evening, hospital administrators were haggling over film rights, and Patrick Steptoe was nursing a tremendous headache. So he slipped away and drove home for a brief interlude to spend a few hours at the piano. Piano playing was his favorite hobby, a holdover from the days when Steptoe earned pocket money working Saturday matinees at the Palace Cinema, offering musical accompaniment to the silent films of Rudolph Valentino and Harold Lloyd.

When he returned with Sheena about eight, his normal hour for making rounds, Steptoe was feeling better. He began to assemble the team needed to perform a cesarean that night — nurses, an anesthesiologist, a pediatrician, Robert Edwards, and Edwards's assistant and traveling companion for the previous five years, Jean Purdy.

At eleven o'clock, the Ministry of Health and the hospital administrators were still arguing over who would own the film, the hospital or the BBC, and who would profit from licensing it to TV outlets around the world. Steptoe announced that if they didn't come to an agreement and sign a contract within five minutes, the whole argument would be moot, as he would allow no photographers into the operating room at all. Two minutes later all the documents were signed.

And so it was that at 11:30 on the night of Tuesday, July 25, 1978, a small team assembled in the delivery room of Oldham General Hospital to attend the birth of the world's first test tube baby. Patrick Steptoe held Lesley's hand for a few moments, said a silent prayer, and listened one last time to the baby's heart, 152 loud, sweet beats per minute. At thirteen minutes before midnight, out came a baby girl, as scrunched and squidgy as any other newborn. She weighed five pounds twelve ounces, and when she took her first breath in the brave new world of which she was soon to be a vibrant, writhing, perfectly shaped icon, little Louise Joy Brown opened her bright pink slash of a mouth and howled.

15
Normality

> Knowledge for the sake of understanding, not merely to prevail, that is the essence of our being. None can define its limits, or set its ultimate boundaries.
>
> — Vannevar Bush, *Science Is Not Enough* (1967)

The worldwide welcome for baby Louise was astonishing. IT'S A GIRL! hailed the New York City tabloids. HERE SHE IS! the London papers announced.

But the acclaim was not universal; the test tube baby's very existence still made lots of people uneasy. Scientists took pains to point out that IVF remained risky, that one apparent success did not mean that this technology could suddenly be used wholesale and without a second thought. "The normality of the offspring is still somewhat in question," Benjamin Brackett of the University of Pennsylvania told a *Washington Post* reporter the day after Louise was born. Brackett was voicing the same concerns Vande Wiele might have if he hadn't been constrained by the lawsuit being argued against him at that very moment. Both men were skeptical, not so much about Louise's normality as about that of the test tube babies who might come after her. A single success, said Brackett, shouldn't lead scientists to leap ahead blithely into in vitro experimentation; there was still much room for error.

"We really don't know the effect of the mechanical and chemical manipulation of an egg prior to fertilization," he said. He added that he personally had no interest in trying to make test tube babies; even though he had successfully created at least one hundred normal, healthy IVF rabbits, Brackett still considered it premature to move on to human beings.

Theologians too were quick to point out that one success did not a revolution make. A few days after Louise's birth, the *New York Daily News* invited representatives of three major religions to a roundtable discussion about IVF. A Protestant minister applauded it "because it expands our knowledge — and knowledge is a gift of God. I think, in this case, what science has done is good because it enhances life, enabling an infertile couple to procreate." A Catholic priest decried it "because we are separating the two aspects of sexuality, the love and the procreative." And a Jewish rabbi disagreed with his fellow clerics' tendency to judge something as ethically acceptable only if it is "natural." "Marriage, for example, is not a natural act," he told the *Daily News*. "Animals don't get married. So humans have already improved on nature . . . In this case, we are improving upon nature to help carry out the religious value of having children."

The debate about whether IVF should proceed just because it had succeeded once was much like earlier debates about other radical new technologies. Emerging technologies, especially those that mimic functions we take to be central to our definitions of life and death and of what makes us unique and emphatically human, often seem gruesome or barbaric in prospect, filled with technical impossibilities or ethical conundrums. Blood transfusion, organ transplantation, mechanical respirators, and artificial insemination — all were greeted with suspicion at first. And then, as soon as these procedures have been done a few times without the sky caving in, the objections tend to fade away.

This was the pattern recognized by a committee of the National Academy of Sciences in their report *Assessing Biomedical Technologies,* which was published around this time. During his tenure at the academy, Leon Kass served as executive secretary of the committee, which focused on what was coming to be known as "technology assessment," a method designed to predict the political, social, economic, and ethical implications of emerging biomedical technologies before they became entrenched. "The initial reactions to a given use of the technology might be very different from later reactions, should it become familiar over years of general use," the Kass committee wrote. Sometimes the change in society's reactions to a new technology is just one of those things; attitudes have a habit of changing. "People may think and feel differently in the future about marriage, procreation, or kinship and the biological family," the committee wrote, "and it is, therefore, risky to predict how they might react to some of the future technological prospects."

IVF, then, was much like other biomedical technologies in its gradual acceptance by a society that at first rejected it. But in many ways IVF was unique. As it toggled so disarmingly between the supremely natural and the supremely artificial, in vitro fertilization straddled another cherished boundary as well: the one between the public and the intimate. The most private, intense moments between a woman and a man were laid bare, through IVF, for lab assault and analysis. Scientists peered into sex cells, the heart of a species' connection to its own future, and they tugged at those cells, messed about with them, and put them back. They became a meddlesome third party in what was, until the advent of their ability to intrude, an eternal and sacrosanct duet.

When Howard and Georgeanna Jones went to Norfolk in July 1978, they intended primarily to help round out the ob-gyn department at Eastern Virginia Medical School. At the time the entire department consisted of just three people: the two of them and Mason Andrews, fifty-nine, their old friend and the man who had recruited them. Beyond that the Joneses thought they might like to create a lab similar to the one they'd had at Hopkins, where research in reproductive endocrinology could occur side by side with treatment, each part enhancing the other. Those pursuits might not put the school on the map, but the combination of solid, respectable science and good patient care would meet a clear and insatiable demand.

But then fate handed them a cue card. On their first day in Norfolk, Louise Brown was born.

"Can a test tube baby be created in the United States?" a Norfolk reporter asked Howard Jones. They were sitting in the middle of the Joneses' new living room, perched on packing crates because the furniture had not yet arrived. "Do you think it could happen here," she asked, "maybe even right here, in Norfolk?"

"I don't see why not," Jones said, assuming that the reporter's off-the-cuff question deserved an off-the-cuff answer.

"What would it take?" she asked.

That one was easy: "Money."

To everyone's amazement, the money suddenly arrived. The day after the newspaper article appeared with Jones's flip plea for support, a Norfolk grande dame called Howard and Georgeanna with a surprising offer. The woman had been a patient of theirs in Baltimore. After years of infertility, she had finally gotten pregnant when Georgeanna came up with the right combination of hormones to induce ovulation. Nine months later the woman was the ecstatic mother of a healthy baby girl, whom she named Georgia in Jones's honor. Now she wanted to express her gratitude in a more concrete way.

Would Georgeanna and her husband accept an anonymous gift to open an IVF clinic in Norfolk? she asked. How much, exactly, would they need?

16

PROMETHEUS UNBOUND

> It is frequently the tragedy of the great artist, as it is of the great scientist, that he frightens the ordinary man.
>
> — LOREN EISELEY, *The Night Country* (1971)

THE DAY LOUISE BROWN was born was the seventh day of the Del-Zio trial. After that, nothing in the courtroom could be quite the same. It became all but impossible for the jurors to turn back to the mindset of 1973 when people were absolutely sure that test tube babies would be deformed and not quite human; that even if they were physically normal they would be crippled psychologically by the knowledge of their bizarre beginnings; that their existence represented a capability that mankind had no right to possess. After that one successful birth, scientists and theologians sitting in judgment of in vitro fertilization might have been uncertain still about whether it was a good idea, but for the general public — including the people on the Del-Zio jury — once baby Louise arrived, the case was closed. The moment she came out so pretty and pink, IVF had proved its worth; test tube babies could be much wanted and much loved, just like other babies, and not doomed to be monsters after all.

It would hardly work, then, for the defense attorneys to claim that Vande Wiele was only doing what any responsible physician would have done, keeping an aberration from being grown for, implanted in, and born to poor trusting Doris Del-Zio. They needed a different line of argument, and they grabbed it the very day after Louise's birth, when a man not even named in the lawsuit took the stand and became, in many ways, the focus of the trial. It all turned on the reputation of Landrum Shettles, the scientist who had tried to beat out everybody else in the race to create a test tube baby.

Who was Shettles, the attorneys on both sides implicitly asked: a visionary or a madman? If he was a skilled embryologist, as plaintiffs' attorney Michael Dennis maintained, the Del-Zios were right to believe he could help them, right to believe that Vande Wiele had essentially committed infanticide. If he was an incompetent kook, as John Bower and the other defense lawyers said, then all that Vande Wiele had done — acting in his capacity as department head, maintainer of organizational discipline, and guardian of Doris Del-Zio's health — was pull the plug on a lifeless soup.

"We do not concede that anything written, proposed, discussed by Dr. Shettles in any publication of his was bona fide, truthful, scientifically acceptable, or accurate," Bower said bluntly. He and his colleagues brought in witnesses to testify that in many cases Shettles had actually used the same photographs again and again, turning them to a different angle, printing them upside down, giving them captions that pretended that the photos were of something new.

Richard Blandau gave some of the most damaging testimony in this regard. A professor of human embryology at the University of Washington in Seattle, Blandau said he was familiar with Shettles's description, in the 1973 *Journal of Reproductive Medicine,* of the embryo transfer he said he had attempted a decade before. Under questioning from Bower, he pointed specifically to that article's Figure Two, labeled as a photograph of a "living human blastocyst five days after fertilization in vitro" and said to have been taken the previous year. Bower passed around copies of Figure Two to the jury. Then he pulled out a big red book in which, he said, an article was illustrated by a photograph that would prove to be relevant. The photograph in the red book, which was published in 1955, was of a human egg at the thirty-two-cell morula stage. It showed, as Blandau explained by reading the caption, an "early segmentation cavity" with "granules in the zona pellucida."

"Now, sir," said Bower, "I want you to take this photograph [Figure Two] and the red book and tilt it around slightly . . . Now, as you have tilted it around, do you have an opinion as to whether or not this photograph [in the red book] and the one that the jury has [from the *Journal of Reproductive Medicine*] are the same photograph?"

"In my opinion they are identical," Blandau said.

"Now this red book from which this photograph was testified by you is what — will you please describe it?"

"This book is the *Journal of Fertility and Sterility,* and it is a well recognized journal."

"What is the article in which the photograph appears?" Bower asked. "Will you read the title?"

"The article is entitled 'The Morula Stage of Human Ovum Developed in Vitro.' "

"And will you please tell us who is the author?"

Blandau's answer, despite the dramatic buildup, probably surprised no one: "The author is Landrum B. Shettles."

At Bower's request, Blandau walked up to the jury box with the two photos, one from 1955 and the other from 1972, and pointed out features common to both. "These two cells here are identical," he said. "All of these various components of this object are absolutely identical in shape and form and position . . . Here is the object turned in this direction. This is the object and it is exactly the same, all the cellular components . . . These little objects are exactly the same in this one. In my view these are identical." Clearly pleased with his exegesis, he resumed his place on the witness stand.

Bower gave him a moment. "Doctor," he asked, "is it possible for a scientist to have taken that photograph in 1972 as part of a bona fide experiment when the same photograph, except tilted sideways, appears in 1955?"

"I would say it would be impossible."

If Blandau was right, then Shettles was a fraud who tried to get away with recycling his own material and misrepresenting it as well.

Despite these aspersions, the plaintiffs' lawyer, Michael Dennis, was convinced that Shettles was not only a good scientist, but a great one. He believed Shettles was brilliant and that his odd personality quirks — his habit of sleeping in the hospital instead of in his apartment, his collection of antique clocks all set at different angles and to different times, his difficulty holding a coherent conversation or looking someone directly in the eye — were idiosyncrasies that should be taken as signs of genius, nothing more and nothing less.

During the lunch break on Friday, July 28, John Bower ducked into the men's room. It was ten minutes past one, in the middle of Shettles's third day on the witness stand, and court was due back in session at two o'clock. Glancing out the window, which overlooked Foley Square, the defense lawyer spotted Shettles standing in the middle of a group of reporters. A spotlight shone on his big bald head as he gave an impromptu press conference. Just one hour earlier, Michael Dennis had explicitly told Shettles not to talk to the press.

Shettles spent seven long days on the witness stand. He had had to move back to New York temporarily from Vermont, a five-hour drive away. (After leaving Columbia, he had taken a job at a small hospital in Randolph.) His testimony was often difficult to understand or even to hear; Shettles talked quickly, mumbled, and used complicated scientific jargon and run-on sentences that befuddled most of those in the courtroom, including the judge. "What does cleavage mean, Dr. Shettles?" Judge Stewart asked at one point, only to be rewarded with a stream-of-consciousness monologue about the egg pronucleus and the sperm head and how "when those actually unite, we have the new life genetically, psychologically." The judge occasionally asked Shettles to slow down, but it didn't help. Along with most of those on the jury, the judge stumbled out of the courtroom each evening that week, not quite sure how the parts of Shettles's testimony all fit together.

Shettles tried to explain why he was absolutely certain that Doris would have been in no danger of infection had he been allowed to implant the mixture into her womb. In his hurried mumble, his Mississippi accent growing more pronounced as his excitement mounted, Shettles offered a lengthy explanation that was heartfelt but not really clear. "Each egg would have been isolated," he said in response to a question from Michael Dennis, "the developing egg into the blastocyst under direct microscopic control in the small tube. It would have been placed into small — about the equivalent of one drop or two drops of the fluid, follicular fluid, sterile. Everything controlled sterile and it would have been introduced by Dr. Sweeney with the catheter technique, the catheter syringe sterile. It is the same technique that all other investigators, Edwards and Steptoe, have used to do their studies, the implantation."

The defense attorneys worked hard to make Shettles look like an incompetent scientist. Presbyterian Hospital's lawyer, James Furey, spent a lot of time questioning him about his record-keeping habits, establishing quickly that Shettles never wrote down anything about his lab procedures and rarely reported on his progress in peer-reviewed scientific journals. The kindest way to interpret this would have been that Shettles was a lone wolf of an investigator, a man so entangled in his interior world that he saw no need to share what he knew with anybody else. The most cynical way would have been that Shettles never wrote these things down because he never did them in the first place. The truth probably lay somewhere in between.

When Shettles testified about having implanted a blastocyst into the uterus of a woman scheduled two days later for a hysterectomy — that landmark 1962 accomplishment that did not come to light until nearly a decade later — Furey asked repeatedly what aspects of the procedure he had recorded. He didn't write down in his surgical report on the first woman, the egg donor, that he had retrieved her ova, Shettles said. Nor did he note in his office records that he had inserted the first woman's egg, five days after it was fertilized with her husband's sperm, into the uterus of the second woman. Why not? Furey asked.

"Because I discussed the situation with her and her husband," said Shettles, "and she knew what I was going to do and she said, 'Well, make some use of the womb or uterus before it is taken out — it might help other people.'"

"Well, that is very laudable, Doctor," said Furey, a slight, stoop-shouldered man who barely seemed physically capable of hammering away at Shettles for so long a stretch. "But don't you ordinarily, when you do a procedure on a patient, make a record of it in your office?"

"In this instance, having the confidence of the couple I did not make any notation," said Shettles, "because I found that nothing takes the place of the rapport, the respect, the confidence of a patient with a doctor."

Furey worked up some sarcasm here. "Was it generally your custom," he asked, "when you have a patient with a great deal of confidence in you and you have rapport with the patient, not to write anything down?" Then he got to the heart of his line of questioning: "Did you consider, Doctor, that this was possibly an historical event? . . . The reason you were doing this, Doctor, was to expand the store of medical knowledge, is that right?"

"That is right," Shettles said, "for the benefit of mankind, but not a selfish one whatsoever."

"Absolutely, Doctor. But wouldn't it have been a good idea for you to write that down so you could remember what you did?"

The lawyers all seemed to love the media spotlight. As much as they criticized the other side for talking too much to the press, they couldn't seem to resist talking themselves when a reporter tossed a question their way. Day after day on the way in to court, at least one lawyer would tarry long enough on the courthouse steps to say something that was heard on the evening news. Day after day Judge Stewart found a reason to admonish one of them for some intemperate remark broadcast the night before. Whenever a lawyer saw a microphone, it seemed, he talked into it.

John Bower, for instance, made a comment one day to a pretty blond television reporter that caused trouble for himself and his client. It's crazy, he told her, apparently caught off guard by her flirtatious manner, for anyone to think "this goo" in Shettles's test tube was even an embryo, much less a baby. When the word "goo" was repeated on the evening news and in the morning papers, Judge Stewart was furious, and told Bower so.

Dennis tried to cash in on Bower's misdeed by siding with the judge. "I don't think you should even refer extracurricularly to 'goo' or 'embryo,'" he told Bower in the judge's chambers, relishing his rare moment of seeming to be the more restrained of the two.

Judge Stewart watched the two attorneys with disapproval, then turned his attention to Furey, the hospital's lawyer, who on NBC the night before had called the case an outrage.

"That is not a correct quotation," Furey explained. "When I was on the steps yesterday, they said to me, 'Mrs. Del-Zio says Dr. Vande Wiele killed her baby. What do you think of that statement?' I said, 'I think it is outrageous.'"

"You think that was a proper statement to make to the press?" asked the judge.

"No," said Furey, who was less confrontational in the courtroom than the other attorneys. "I am sorry I said it."

To which Judge Stewart wearily replied: "I think it is time for us to stop creating opportunities in which you can express your sorrow."

The lawyers could easily have avoided all these showdowns in the judge's chambers. They could have shouted a brusque "No comment" and kept on walking when reporters accosted them outside the courthouse. But each of them, and their clients as well, understood the power of the press attention, knowing that their real battle was being waged not in the courtroom but out on the larger stage of public opinion. It was all part of the legal strategy — and was also part of something that mattered even more to Shettles and Vande Wiele, the great push-and-pull fandango of introducing a new scientific idea to a recalcitrant society.

Raymond Vande Wiele on the witness stand was impeccable; a solid, good-looking man with a fleshy face and wavy gray hair, he was always dressed carefully in a three-piece summer suit. His credentials were impeccable, too — and for the first few hours of his testimony, his credentials were all he talked about. He laid out carefully, year by year, the series of impressive places where he had done his work. Undergraduate degree from Catholic University in Louvain, Belgium. Medical degree from the same university, in 1947. Internship and residency in Louvain, postdoctoral fellowship at Universität des Klinik in Vienna, a Fulbright fellowship in reproductive endocrinology at Columbia in 1953, his first appointment in America. After a one-year fellowship at Yale in 1955, he returned to Columbia. And there he remained, moving up from instructor in 1956, to full professor in 1966, to chairman of the department of obstetrics and gynecology in 1971.

Under questioning from his attorney, Stephen O'Leary, Vande Wiele pointed out that he was a member of "fifteen or twenty" professional societies, two of which he mentioned by name. One, the Association of Professors of Gynecology and Obstetrics, had only eighty members nationwide; "therefore," Vande Wiele said, "its selection is extremely exclusive." The other, the Society of Clinical Investigation, was "a very prestigious society, virtually limited to people who are in internal medicine. People in internal medicine have a tendency to think they are better than the other specialties. In fact, if I am correct, I am quite sure, they have over the years only admitted three obstetricians, and I am one of them."

Besides his "exclusive" society memberships, Vande Wiele testified, he was frequently asked to serve as a visiting professor at medical schools all over the world. In 1978, which was only half over, he had already traveled in that capacity to India in January, Israel in April, and, just the previous month, to Japan.

So there he was, the suave professor with an international clientele, his accent a faint echo of Europe's antique aristocracies, his ties always knotted just so, trying to explain why he had uncorked a test tube five years earlier to stop an experiment he thought should not proceed. He took command of his attorney's line of questioning, answering not only in carefully parsed paragraphs but with subheads and annotations. He had five reasons for stopping the experiment, he said, and proceeded to enumerate them: "The first reason was that I felt this was a danger to the patient in whom the procedure was going to be carried out.

"Two, I thought there were ethical and moral objections to this procedure.

"Three, I had been specifically told by my superiors to stop this experiment.

"Four, it was in clear violation with the rules of the Department of Obstetrics and Gynecology and with the assurance filed by the College of Physicians and Surgeons [promising to abide by the regulations of NIH].

"And, finally, I felt that it was carried out by individuals who had no competence to carry out such a procedure."

Stephen O'Leary was a young man without much courtroom experience, but he was savvy enough to see that Vande Wiele was coming off as rigid and autocratic, someone who relied on an inflexible reading of the rules. Yet he believed that the impression of aloofness didn't matter. Vande Wiele did not have to be a cuddly figure. All he had to be was confident enough — and better qualified than Shettles — to know which scientific interventions made sense and which ones did not.

Michael Dennis, the more experienced plaintiffs' attorney, believed Vande Wiele's confidence could be persuasive — but he also thought it was mostly a veneer. His intention was to try to find a way to crack that veneer. Or maybe cracking it wasn't the point; maybe all Dennis needed to do was highlight it, polish it to a gleam, twist it in the light until it reflected just what Dennis wanted it to reflect.

"I am not a coward," said Vande Wiele under Dennis's cross-examination on his second day of testimony. "I don't run away from decisions."

"In other words," said Dennis, "you consulted your superiors, is that right?"

"That's correct."

"And they ordered you, is that right?"

"That's correct."

"And all you did was follow orders, is that right?"

The defense attorneys saw where this was going. "I object," said John Bower. "The way he phrases it is obviously designed for another purpose."

"Overruled," said Judge Stewart. But Bower's objection hung in the air. Michael Dennis was insinuating — unfairly and irrelevantly, in Bower's view — some sort of Nazilike behavior on Vande Wiele's part. Why else would Dennis also ask about the exact dates of Hitler's occupation of Belgium and about the Nuremberg code, in particular the reasoning behind its prohibitions against certain forms of human experimentation? Dennis's logic was muddy and his execution awkward, but one thing was clear: he was hoping somehow to entwine in the jury's mind this haughty European with the worst outrages of Nazi brutality. In private he referred to Vande Wiele as "Wilhelm," making sure to pronounce the first letter with a Germanic "V."

Whether or not he had been following orders, Vande Wiele patiently listed his reasons for uncorking that test tube again and again during his three days on the witness stand. The jurors could probably tick them off in their sleep: danger to the patient, ethical objections, orders from superiors, procedural irregularities, incompetence of the investigators.

One of Vande Wiele's biggest objections to Shettles's experiment was that in 1973 it was premature to do IVF on humans. The defense called several expert witnesses to bolster this point of view. Both Richard Blandau of Seattle and another defense witness, Luigi Mastroianni of the University of Pennsylvania, believed more animal work was necessary, even at the time of the trial, before IVF could fully be embraced. Who is to say, they testified, that Louise Brown's apparent health wasn't some sort of lucky fluke?

"Is it your view, Doctor, that IVF research on monkeys and rabbits [is] useful to IVF research in humans?" Dennis asked Blandau in cross-examination.

"I think they are essential," said Blandau.

But what about the argument by Mastroianni, who had been quoted in *Ob/Gyn News* as saying, "We can find out all about monkey and rabbit fertilization by working on monkeys and rabbits, but not learn anything about human fertilization until we work on humans." Do you agree with that sentiment? Dennis asked.

Blandau demurred. "Obviously if one works on the human, then you have a situation which is human," he said, sounding like Humpty Dumpty in *Through the Looking Glass.* "If you work on rabbits, you have a situation which is rabbit."

"Precisely."

"But both of them are absolutely interchangeable."

"That is your view?"

"Absolutely."

The truth was, neither Blandau nor anyone else at the time knew exactly what animal research could reveal about humans or exactly how much human research would be needed before IVF could be transformed from a bold experiment to a conventional medical therapy.

Another of Vande Wiele's objections to the Sweeney-Shettles experiment was that he believed it violated NIH regulations. He said there was a government ban on using federal funds to do research on human embryos. And if a single procedure failed to accord with the letter of assurance that Columbia had filed with NIH, then, in Vande Wiele's reading, every one of the medical school's federal grants could be terminated. He estimated the potential in lost research support for Columbia in the millions of dollars.

While the matter of what the government did or did not allow should have been easy to resolve, it turned out not to be. Vande Wiele tried to make it appear otherwise in his testimony, but this issue was a controversial one. Did the government allow research on IVF in 1973? in 1978? The answer apparently was one on which reasonable men could — and, most heartily, did — disagree.

The Ethics Advisory Board of HEW, which was holding hearings on IVF at precisely the moment Louise Brown was born, tried to weigh in on the matter. Although fact-finding wasn't strictly part of the board's mandate from Secretary Califano, two staff members flew to England to visit Steptoe's hospital within days of Louise's birth, eager to get details about how the egg was fertilized and transferred and, possibly, insight into whether something similar should be done in the United States. They found Steptoe surprisingly uncooperative. One of the emissaries, staff director Charles McCarthy, said Steptoe was cagey about revealing details of the work, offering only that a scientific account would be published before the end of the year. The other, staff counsel Philip Halpern, said that some of the other scientists they met in Britain weren't especially impressed with Edwards and Steptoe's achievement. Many saw it "not as a scientific breakthrough," he said, "but as a technological one." And, he added, many believed the pair had "left out several important scientific steps along the way, and therefore contributed little to our store of knowledge."

Edwards and Steptoe's reticence to offer details about their procedure left the door open for some pretty wild speculation about how they had made IVF finally work for Lesley Brown. One rumor was that Steptoe had retrieved Lesley's egg in Oldham, while Edwards had fertilized it with John Brown's sperm in Cambridge, 250 miles away. It sounded a bit like the geography of the Shettles-Sweeney collaboration, writ large. Because of the distance, according to the rumor, Edwards and Steptoe had resorted to an interesting conveyor: the Brown ovum had made its way to Cambridge and the Brown blastocyst had made its way back to Oldham safely housed in the uterus of a rabbit. The rumor was so delicious that one newspaper carried a cartoon of Steptoe driving around town in his Mercedes with a huge female rabbit sitting primly in the passenger seat.

The fierce publicity surrounding the case of *Del-Zio v. Vande Wiele et al.* was more than anyone had bargained for. Standing-room-only was typical, the room packed with journalists scratching at their skinny notepads, courtroom artists flourishing their pastels, spectators drawn in by the stories on the nightly news. It was so noisy that some witnesses could barely be heard; someone needed to find a microphone for Doris Del-Zio, her soft voice was so overpowered by the background hum. Every day the jurors watched the hubbub in the courtroom, the scribbling journalists, the focused artists, the attorneys leaping to their feet. Every evening they passed reporters shoving microphones toward witnesses on the courthouse steps. And every night they could, if they chose, fall asleep to Johnny Carson making test tube baby jokes on NBC.

"They're having the biggest trouble with the test tube baby at feeding time," Carson said one night shortly after Louise Brown was born. "The mother holds up the bottle, the baby says 'Da da,' and tries to crawl inside." Another night, another groaner: "Yesterday scientists at Stanford Medical Research laboratories announced that they've developed a birth control device for test tubes." And: "Will this never end? The other day I read about a test tube baby who married a clone and got pregnant through artificial insemination." The jokes were not of the highest caliber, but as social commentary they were potent indeed. Louise Brown's cameo appearances in the *Tonight Show* monologue night after night was an indication of a certain kind of fame. And within days of her birth, an astounding 98 percent of Americans, according to a Gallup poll, said they had heard of in vitro fertilization, and 60 percent could give a rough approximation of how it worked.

The court proceedings included sufficient spectacle to keep the audience continually titillated. This was, after all, a trial about reproduction, with plenty of chances to talk about eggs and sperm, genitals, masturbation, all sorts of not-quite-respectable topics. There were also juicy allusions to professional jealousies, personal vendettas, Nazi allegiances, and medical incompetence. There was an Irish doctor in a yarmulke with a lewd vocabulary, a southern doctor who talked so fast he could barely be understood, a Belgian doctor who treated the courtroom like his own personal lecture hall, and an anguished and aggrieved woman who often burst into tears.

But for all the theatrics, the trial ultimately revolved around two essential questions: whether the stuff in Shettles's test tube was or was not a baby and whether it should or should not have been destroyed. The defense's main argument, in their effort to prove that Vande Wiele was right in terminating the experiment, was that Shettles's mixture was not a zygote at all but a contaminated brew that would have endangered Doris Del-Zio's health had it been inserted into her uterus. That point was made most forcefully by Vande Wiele's colleague at Columbia, the woman known as "Mrs. Egg," Georgianna Jagiello.

A professor of reproductive endocrinology, Jagiello was widely recognized as one of the world's leading experts on the human ovum. She believed that Shettles's attempt at IVF was messy and misguided, and that if he had been permitted to implant the specimen sitting in

Mary Parshley's incubator, his patient's life would have been in danger. She based her conviction, she said, on her own experience at Guy's Hospital in London in the 1960s, which at the time was one of the world's leading centers for the study of the external fertilization of human eggs. According to her testimony, it was at Guy's that she learned that test tubes didn't work as well for fertilization as flat culture dishes, which exposed more of the solution to carbon dioxide and thereby kept the pH at the proper level.

Shettles had used a test tube.

It was at Guy's that she learned that black test tube stoppers didn't work as well as red stoppers made of natural products, because the synthetics used in the black stoppers tended to emit fumes that killed human eggs.

Shettles had used a black stopper.

And it was at Guy's that she learned that unless an incubator had a mechanism for regulating the flow of carbon dioxide, the solution could drift toward the acidic, which could also kill the fragile human eggs.

Shettles had used an incubator with no such mechanism.

"We learned very early on," Jagiello added, "that blood is to be avoided at all costs because it kills the ova." Blood was also "a well known culture medium for bacteria — they grow very well in it." Shettles had put lots of things into his test tube along with the egg and two drops of semen — follicular fluid, tubal mucosa, human placental blood serum, uterine scrapings — any of which could easily have contained traces of blood.

Jagiello's testimony was damaging to the Del-Zios. It undermined their assertion that the fertilization had worked, that everything would have gone according to plan had Vande Wiele not intervened, that Vande Wiele had "killed" a growing pre-embryo. Jagiello's testimony made it seem likely that Shettles didn't have a clue what he was doing, and that it would have been a miracle if the stuff in his test tube had ever coalesced into a fertilized, viable zygote.

How disconcerting it must have been for the jurors to listen to Jagiello. Here she was, on the seventeenth day of the trial, tall and steely, with her precise bun and simple string of pearls, contradicting just about every one of the plaintiffs' claims about the events of September 1973. That stuff in the test tube was "my baby," Doris Del-Zio had insisted. Nonsense, said Jagiello; it was a bloody mess with an "unseemly appearance." Vande Wiele was concerned about nothing but his own reputation, Landrum Shettles had said. Wrong, said Jagiello; he was merely trying to protect the health of a woman he had never met. The destruction of the specimen in Mary Parshley's incubator reflected Vande Wiele's callous disregard for the Del-Zios' feelings, said Michael Dennis. No, said Jagiello; it reflected his heroism.

In Jagiello's view Shettles was a renegade and Vande Wiele a saint. Instead of suing him, the Del-Zios should be getting down on their knees to thank him for saving Doris's life.

The arguments finally ended on the afternoon of August 17. The three defense lawyers had tried to paint Doris Del-Zio as a conniving prima donna, to show that Landrum Shettles was a has-been and a clown, and to plant doubt that there was anything viable in that test tube to begin with. The plaintiffs, in turn, had tried to make Raymond Vande Wiele seem like a hot-headed, small-minded bureaucrat, a man jealous of the greater prestige and visibility of his junior colleague.

And as is the way in the American system of adjudication, it was up to the jury to decide which side to believe.

17

VERDICT

> Science knows only one commandment — contribute to science.
>
> — BERTOLT BRECHT, *Life of Galileo* (1939)

WHEN THE DEL-ZIO TRIAL was finally sent to the jury, five long weeks after it began, Judge Charles Stewart spent an unusual amount of time formulating his charge to the jurors. He did so in part because this was one of the highest-profile trials he had presided over in his years on the bench. But he did so, too, because he was aware that although in many ways it seemed like a landmark lawsuit, legally speaking, it wasn't. Stewart did not want the jurors to get too carried away.

The law is one of society's most forceful ways to regulate how people act toward each other and what they owe one another. It is an arbiter of decency and civility, a way of establishing what kind of behavior is acceptable and what is not. But the law is often rendered mute in the face of complex scientific issues of the day, even those that have a legal element. Juries may decide, based on the merits of a particular case, what constitutes medical malpractice, or how much harm from environmental pollutants can be tolerated, or whether a drug or tobacco company is liable for the disease its products cause. But even if a verdict sets precedent or sets the tone for social behavior, juries cannot set science policy. That is an exercise left to Congress, the president, and the proliferation of expert boards and commissions set up to advise them.

This was true in the Del-Zio trial, too. The timing of world events, and the magnitude of the issue itself, might have made it seem otherwise, but this case was not about whether in vitro fertilization was good or bad; not about whether test tube babies were likely to become more common in the future; not about whether infertile couples had an inherent right to children of their own; not about whether IVF was the first step down a slippery slope toward designer babies and wombs for hire.

What it was about, as Judge Stewart put it, was quite specific: whether the "greater weight of the evidence . . . that is, its convincing quality, the weight and the effect that it has on your minds" indicated that Raymond Vande Wiele was guilty of "intentional infliction of emotional distress" upon Doris and John Del-Zio. What the jury was being asked to decide, he said, was whether Vande Wiele exhibited conduct "so extreme, outrageous, and shocking that it exceeded all reasonable bounds of decency." The implications of that decision, in terms of society's attitude toward IVF and where it might lead, would have to be played out in some other place.

Given the frantic media attention generated by the trial and by IVF in general, Judge Stewart told the jurors there was a good chance that the verdict would set the tone for subsequent debates about such matters as when life begins and who has the right to begin it. But he wanted them to remember that their decision would have a direct effect on one group only: the litigants. It would set no legal precedent for how the law should view collections of human cells — as property, as potential life, or as life itself.

The jury of four women and two men deliberated for thirteen hours, beginning on Thursday afternoon, August 17. The foreman, Hugh Lawless, was a lifetime employee of the city's electric utility, Con Edison. In big, fat handwriting, Lawless wrote politely worded questions on a legal pad and passed the long yellow sheets, one by one, to the clerk outside the jury room, asking for a transcript of a witness's testimony or an explanation of a particular point of law. His requests gave clues about what the jurors considered important, what they had trouble remembering, where they disagreed. One of the first notes came out of the jury room just before two o'clock.

> Your honor:
> May we have the hospital records for the September, 1973, operation on Mrs. Del-Zio.
>
> Thank you,
> Hugh Lawless,
> Foreman

At five o'clock he again requested those hospital records.

At 5:48 P.M. came this handwritten note:

> Your honor:
> May we have one or two copies of the charge that was read to us by you.
>
> Thank you,
> Hugh Lawless,
> Foreman

At just after nine that evening, Lawless requested the transcript of Jagiello's testimony regarding the phone call to Vande Wiele, along with Vande Wiele's testimony. The jury broke off their conversation shortly before ten, most of them too tired to think straight. Rooms had been reserved for them at a nearby hotel, and dinner was provided.

When they resumed deliberations the next morning, they still needed more information. At 11:08, Lawless sent out another note.

> Your honor:
> The jury would appreciate clarification of the last paragraph on page 18 of your charge to it which appears to be in direct conflict

with Dr. Jagiello's testimony. As we understand it, Dr. Vande Wiele asked Dr. Jagiello to have Dr. Toran-Allerand bring the test tube to his office in the same conversation in which she informed him of its existence.

Thank you,
Hugh Lawless,
Foreman

By now more than twenty hours had passed since the jury first retired, and the waiting was unbearable, at least for Doris. Nervous and edgy, she just couldn't sit still. She paced the hallway outside the courtroom until she practically wore out the floor. Raymond Vande Wiele had no such problem. He sat quietly during the entire wait, reading or working on crossword puzzles. He and the Del-Zios never spoke.

At 3:14 P.M. the jurors sent out a note with one more question — a more technical one this time, which made it seem that they might be closing in on a decision:

Your honor:
If a claim were to be awarded based on the first paragraph of page 41 of your charge to the jury, is it also necessary to accept the premise enunciated in Paragraph 3 on the same page. We would appreciate your clarification on this point.

Thank you,
Hugh Lawless,
Foreman

One woman on the jury didn't trust the Del-Zios, didn't believe a word they said, didn't want to give them a penny. The other jurors had different ideas, some of them arguing that even if the amount that the plaintiffs were asking for was too high, they at least should get a settlement that would cover their legal fees.

After a few more hours of haggling, the jury reached a consensus on the second evening. At 7:30 P.M., the clerk called back all the parties involved — John and Doris Del-Zio, Raymond Vande Wiele, and their assorted attorneys — to hear the verdict. It did not take long for everyone to assemble in the third-floor courtroom.

The air was artificially chilled and naturally fraught, as it had been for most of the trial. But, at least for now, the sniping and accusations had stopped, and the Del-Zios sat quietly beside their lawyer, while Raymond Vande Wiele and a bunch of attorneys sat just as quietly across the aisle.

Much had changed since the trial's beginning five weeks before. The summer had ebbed away. The world's first test tube baby had been born, healthy and normal, in England, proving that a life created in the lab could look like any other life. The jury foreman, taking daily lunch breaks in Chinatown and Little Italy, had gained twenty pounds. Landrum Shettles had left his new home in Vermont to hang around New York, John and Doris Del-Zio had shuttled to a relative's house in Queens, and Raymond Vande Wiele had lost a summer that he had hoped to spend traveling abroad.

Lawless and his fellow jurors did their best to mete out what they saw as justice. They found Raymond Vande Wiele liable for "atrocious" behavior that was "utterly intolerable in a civilized community," but the payments they required were just 5 percent of the $1 million Doris Del-Zio had asked for: $12,500 each from the hospital and the university and $25,000 from Vande Wiele. As for John Del-Zio, who had asked for $500,000 for the loss of his wife's consort, the jury found that he was injured but that the damages could be covered with an award from each of the three defendants of just one dollar.

The awards, especially the three dollars to John, were a stinging little slap on the wrist for the Del-Zios, with the message that even though they could be said to have won the case, their complaints were out of proportion to their actual travails.

Even so, Vande Wiele was hurt, shocked, distraught; he had

thought he would be vindicated. To clear his name, he immediately filed a motion to have the verdict set aside. In September the motion was denied, and his lawyer advised Vande Wiele not to appeal. Still steaming, he was forced to turn his back on this frightful episode. But he would never, for the rest of his unexpectedly brief lifetime, forget its bitter lessons.

Part Four

Not Meant to Be Known

18

Right to Life

> There is a lurking fear that some things are "not meant" to be known, that some inquiries are too dangerous for human beings to make.
>
> — Carl Sagan, *Broca's Brain* (1979)

Monsters. genetic defectives. Chromosomally damaged beasts erupting from an unnatural egg-and-sperm tango in a petri dish. The prospect of abnormal babies was, without question, the scariest and most titillating thing about in vitro fertilization. Yet some of these fears fell away amazingly quickly. A Harris poll taken in August 1978, soon after Louise Brown's birth and right after the verdict in the Del-Zio "test tube death trial," showed a surprising acceptance of IVF. More than half of those polled said they would use IVF themselves if they were married and couldn't have a child. Younger people were more willing than older people to consider turning to IVF; two-thirds of adults under thirty said they would try it if they had to. Overall, 60 percent of Americans in August 1978 — a surprising majority, given the fractious nature of the times and the revolutionary nature of the procedure — thought IVF should be available for anyone who needed it and wanted it.

But some Americans still harbored deep and passionately held reservations about test tube babies. It made scientific sense that there would in fact be aberrations, that all those laboratory manipulations might lead to genetic derangements that could result, at least occasionally, in the monsters of our collective nightmare. Maybe the world's first test tube baby was just beginner's luck.

These concerns were not limited to the uninformed masses and the readers of supermarket tabloids. The scientists expressed them too. Even Edwards and Steptoe had to admit the possibility that Louise Brown, cooing so sweetly in her grateful parents' arms, was an anomaly. That's certainly how it started to look when the British scientists failed to have a second IVF birth for several months. Their second test tube baby, also healthy, was not born until January 4, 1979, more than five months after Louise. In the interim were two failed pregnancies, one a miscarriage at twenty weeks after the mother went on a hiking holiday, the other a genetic abnormality known as a triploidy, a fetus with too many chromosomes to support life. The miscarriage was a setback, but it was the triploid pregnancy that struck terror in the hearts of the Oldham team, which saw it as possibly undoing all the progress that had gone before. The triploid embryo was, at last, the dreaded beast.

A triploidy results when two sperm inseminate a single egg. During natural conception, the egg undergoes a chemical change the instant the first sperm penetrates, shutting down entry to all subsequent sperm. But no one could be sure that the same signal would switch on in a petri dish. Edwards had worried from the start that "artificial conditions pertaining in the laboratory might, in some mysterious way, interfere with this chemical alteration in the egg membranes." Or perhaps the presence of so many spermatozoa — orders of magnitude more than the number that could swim up the fallopian tube on their own — would swamp the egg's shutdown mechanism. A two-sperm fertilization is a death sentence: it means that instead of the zygote having the normal complement of forty-six chromosomes (twenty-three from the egg, twenty-three from the sperm), it has sixty-nine: twenty-three from the egg, twenty-three from the first sperm, and twenty-three from the second sperm. Nature can do nothing with such a thing but expel it.

Despite the risk of triploidy and other aberrations, however, scientists kept making plans to bring IVF to more and more infertile couples in Great Britain; in India, where the world's second test tube baby was reported to have been born on October 3, 1978; in Australia, where a test tube baby clinic opened in Melbourne even before Louise Brown was born; and, stimulated by an infusion of unsolicited donations, in the United States. It was the American clinic that caused the most startling outbursts, primarily from religious fundamentalists who thought that the "right to life" — the era's slogan for antiabortion sentiments — did not include the right to create life anew.

Howard and Georgeanna Jones hadn't set out to become gray-haired pioneers, hadn't meant to take on an entire political movement, hadn't intended to open the first IVF clinic in the United States. But by early 1979 the momentum for such a clinic was palpable. After that first gift from their anonymous benefactress, money kept arriving, all of it earmarked for a new IVF clinic at Eastern Virginia Medical School in Norfolk. Within a few months the Joneses had amassed some $25,000 — enough to buy some lab space and equipment and to get things rolling.

Opening the clinic would put the Joneses at the forefront of one of the most controversial branches of American medicine, but that didn't really scare them. Georgeanna Jones had a certain reticence, but her husband — her opinionated, stentorian, unflappable husband, Howard — had feistiness to spare. Howard Jones always said that he loved a good fight. Which was lucky, because a fight was what he was in for.

Lucky, too, that the medical school administrators also loved a good fight — and relished the opportunity to earn a national reputation, serve a desperate and growing market niche, and make a good deal of money, all in a single grand gesture. This is what made the Joneses' situation at Eastern Virginia in 1979 almost the polar opposite of Shettles's situation at Columbia in 1973. Eastern Virginia was a brash upstart, eager to make a mark and convinced that the way to do that was by testing the limits of frontier science. Columbia was dowdy and austere, cautious about its hard-earned reputation and loath to take a step too far in the direction of change.

The personalities, like the institutional settings, could hardly have been more different. Eastern Virginia's chairman of obstetrics and gynecology, Mason Andrews, was a charismatic risk taker with a flair for the dramatic gesture and a fondness for a particular llama-shaped oversized tie clasp. A popular obstetrician, he ran successfully for his first term as city councilman with a double-entendre campaign slogan that gave some insight into his southern-fried spirit: "Mason Delivers."

His counterpart at Columbia, Raymond Vande Wiele, had none of Andrews's spunkiness and an entirely different brand of charm. And while Vande Wiele could be an insightful and creative scientist, he did not like to take risks — especially when the image of his institution was at stake.

Different, too, were the reputations of the investigators themselves. The Joneses were longtime leaders in reproductive endocrinology, charter members of what Jones dubbed the "IVF Mafia," the handful of people involved in the field in the early and mid-seventies who shared their triumphs and frustrations at a series of exclusive retreats around the world. They were integrated into the IVF community not only intellectually but socially. Before they relocated to Norfolk, their seven-bedroom home in Baltimore had for years been the setting for crabfests and informal get-togethers with international luminaries in reproductive medicine.

Landrum Shettles, on the other hand, was forever an outsider, a researcher out of step with the rhythm of the day, whose social oddities kept him from forging scientific or political alliances. His idiosyncrasies made him unable even to live with his own family — and his longest-term collaboration was not with a scientist at all but with a reporter.

How strange that the geography of the story turned out this way, going against all conventional wisdom, all expectations of the demarcations of popular and intellectual culture. If you envision where you would most likely find an aggressive, boundary-pushing collective of scientists, it might easily be New York — and it would almost certainly not be Norfolk, a town one-tenth the size, semirural in its pace, deeply religious in its attitude.

Yet in New York was Raymond Vande Wiele, who would stop an associate's attempt at the first American IVF, while Norfolk's Mason Andrews almost gleefully stepped right into the quagmire and gave his colleagues every encouragement to do whatever it took. At Columbia protocol was crucial because there was so much to lose, while Eastern Virginia Medical School was making its mark by ignoring the usual rules. The people in Norfolk, unlike those in New York, were edgy and eager, people without much time to lose, and the institution they worked for was quick to respond to the vicissitudes of supply and demand.

In March 1979, just as plans for the Norfolk clinic materialized, the federal government's Ethics Advisory Board issued a formal report declaring IVF ethically acceptable so long as certain conditions were met. According to the board, it was ethical to conduct research on embryos that would not be implanted, "even if the new embryo is then destroyed," as long as the effort was meant to improve the safety or efficacy of IVF or to gain "important scientific information" about the ways human babies start to develop — and as long as the embryo was kept alive for no more than fourteen days. This meant that Pierre Soupart's NIH proposal to create, examine, and destroy some four hundred human IVF embryos had at last jumped through its final hoop.

Not only could Soupart's research go ahead but so, it seemed, could plans for the Norfolk clinic. Even though the clinic would use no federal research funds, the lifting of the IVF moratorium would surely have a salutary effect on the Joneses' efforts. Since the government had given its blessing to basic scientific investigations of IVF, wasn't that a kind of tacit approval of clinical investigations, too? And if the government was willing to support IVF research with substantial sums of money, didn't that indicate that it would have no objections to in vitro fertilization underwritten entirely with private funds?

Still, despite the Ethics Advisory Board report, and despite Mason Andrews's unqualified endorsement, and despite the range and clout of the Joneses' scientific allies, the road to the Norfolk clinic was mined with booby traps. Some of the opposition came from scientists, who thought it was still too soon to introduce IVF into medical practice. Some came from political liberals, who objected to the notion that we are nothing more than the sum of our parts, that humans can be built from the gamete up. But most of the opposition came from the people one would expect to be the biggest opponents: the right-to-lifers, who were as plentiful in the Tidewater region of Virginia as mollusks after a summer storm.

The local chapter of the Virginia Society for Human Life soon went into action after the Joneses announced their plans to open an IVF clinic. Such an undertaking, said executive director Charles Dean, Jr., would inevitably involve flushing "imperfect babies" down a laboratory sink. "There was no proper study of the medical, moral, legal and scientific merits," he told a reporter for the *Washington Post.* "Meanwhile we're charging off into the darkness, and I feel it's a tragedy and a disgrace."

Dean was an extremist, a religious fundamentalist who got more worked up about IVF than did the mainstream branches of Christianity. In New York City, for instance, the Queens Federation of Churches, a Protestant consortium, called in vitro fertilization "very exciting because it expands our knowledge — and knowledge is a gift of God . . . [IVF] is good because it enhances life, enabling an infertile couple to procreate."

Dean might have been extreme, but he was not alone. He was able to rally dozens of other right-to-lifers for marches and protests; to find advocates among state legislators and congressmen; to get friends to write letters to the editor and to the medical commissioner as though they were objecting with one voice. His zealotry, and his network, made him a formidable foe.

For the IVF clinic to start operating, the sponsoring hospital, Norfolk General, required a certificate of need, a document issued by a local department of health and required of any hospital hoping to branch out into a new field of service. The certificate of need originated in the early 1970s, when expenditures for high-tech medical equipment accounted for much of the upward spiral of health-care costs. Having to obtain the document, according to health-care economists in the Carter administration, would keep hospitals from unnecessarily duplicating one another's services. If Hospital A already had an expensive machine like a CAT scanner, Hospital B could be denied a certificate of need for a scanner on the assumption that the community didn't need a second one, that Hospital A's was meeting the community's demand.

Obtaining a certificate of need was an accounting matter, not an occasion for political grandstanding. Applications were usually debated during low-profile meetings of the state health commission, with no outside testimony required. That's what Glenn Mitchell, the administrator of Norfolk General, expected in this case. He also expected the outcome to be in the clinic's favor. Clearly, there was no question of duplication of services. Not only were there no other such clinics in the Norfolk area or elsewhere in Virginia; there were no other such clinics anywhere in the United States. A new building was not needed, and little had to be purchased: the hospital, which was using only four of its six delivery rooms, was offering the Joneses one of the extras for the IVF clinic, as well as a sterilizing room next door that could be converted into a laboratory. Mitchell told Howard Jones not to bother coming to the session of the state health commission in which the IVF clinic's certificate of need was to be considered. He assumed approval would be routine.

But then Charles Dean and his supporters heard that the IVF clinic was to be certified at a quick administrative session in August 1979. The right-to-lifers packed the meeting room at the state medical commissioner's office. "The last thing Virginia needs," said one of them, "is such a laboratory where scientists will be playing God with babies' lives." Frightened by the opposition, the commissioners balked at approving the application as they had intended. Instead, they voted to hold a public hearing to consider the certificate of need issue more openly. And through an oddly poetic twist of timing, the matter was placed on the docket for the afternoon of October 31, 1979 — Halloween.

With its religious roots in All Soul's Day, celebrating the recently departed, Halloween is a time when the borders between the living and the dead are blurred. People dress in black, with ripped fishnet stockings and lips the color of a bruise, or they pretend to be ghosts, skeletons, hideous corpses — anything that stinks of death. Or of death come to life; that, after all, is the essence of the celebration. In the fifth century B.C. the Celts believed that on Halloween the spirits of those who had died the previous year came back to earth looking for bodies to possess. If they managed to inhabit a living body, their souls became immortal; if not, they died for good. Many of the rituals associated with the day, like wearing costumes and making noise, come from the ancient belief that doing so would keep the dead from choosing to take up residence in your particular body.

Tidewater area fundamentalists planned to pack the auditorium for the Halloween hearing just as they had two months before at the medical commissioner's office. They wore no costumes, but they showed the same fervor they had in August, carrying signs with slogans like UNNATURAL CREATION, WHO SHOULD PLAY GOD?

and INCEST IN A TEST TUBE (that last one a real puzzler, making one wonder whose gametes, exactly, they thought were being mixed).

This time, however, the clinic advocates were prepared. They had invited a group of impressive figures to testify on their behalf—not only the expected choir, including gynecology professors from Harvard and Duke, but surprise witnesses like the Episcopal bishop of Virginia and a prominent Norfolk rabbi. And then, at precisely 1:30, a group of medical students from Eastern Virginia Medical School marched out of their classroom building and headed across the street to where the hearing was to be held. When the right-to-lifers arrived by the busload at two o'clock, there were no seats left. All three national television networks were there, as well as the fire marshal, keeping most of the crowd out on the front steps in accordance with fire regulations. The hearing lasted well into trick-or-treat time, not adjourning until nearly eight o'clock.

Two weeks later came another hearing — more sedate than the one on Halloween — after which the state health commission voted unanimously, with one abstention, to approve the certificate of need. This meant that the state medical commissioner, James B. Kenley, was required to approve the clinic. Senator Orrin Hatch, a conservative Republican from Utah with a staunch right-to-life constituency, phoned Kenley directly, asking him to turn down the request despite his board's directive. It's too soon to make so precipitous a decision, said the senator, "on a matter that has the potential to radically alter our view of human creation."

Kenley resisted Hatch's entreaties and resisted too the threat of a lawsuit made by the relentless Charles Dean. He ignored the seven hundred letters opposing the clinic that formed a pile in his office more than twelve inches high. And on January 8, 1980, he approved a certificate of need, as directed by his commissioners. There was, in the final analysis, no regulatory basis on which to deny the request.

A month later, in early February 1980, Dean tried another tactic. He asked the attorney general of Virginia, Marshall Coleman, to impose an injunction against the clinic's opening, on the grounds that the approval of the certificate of need was the result of "collusion" among members of the various review panels. Coleman, while expressing sympathy with the goals of the Virginia Society of Human Life, said he had no legal standing for taking such an action. Dean also turned to the state legislature, finding an ally in Delegate Lawrence D. Pratt, a Republican who represented the staid, affluent Fairfax County suburbs of Washington, D.C. Pratt held hearings on the issue and tried to introduce a bill that would make IVF illegal in Virginia. He said he feared a "supermarket approach" to choosing babies with particular genetic traits. He didn't want to outlaw the clinic, Pratt said; he merely wanted to rein it in, keep it from going beyond its authority. He also wrote anti-IVF editorials. "My investigations to date indicate that the technique of in vitro fertilization is highly experimental with untold potential for future tragedy," he wrote in one such column, in the *Virginian-Pilot.* "The doctors who engineered Louise Brown's birth also succeeded in successfully implanting embryos in two other women. Both fetuses spontaneously aborted and were found to be quite abnormal." He did not say where he had gotten this information, which contradicted what Edwards and Steptoe were saying publicly about their post-Louise experiences.

But Pratt — a man accustomed to running successful direct-mail campaigns for conservative candidates and for his own organization, Gun Owners of America — was up against a stronger tide than he could handle. Growing sentiment in favor of IVF came with a slightly patriotic tinge — it was part of an urge to get the United States back into the international race for scientific supremacy. The British, the Australians, even the Indians were reporting IVF successes and leaving American efforts in the dust. Opening the Norfolk clinic was becoming a kind of biological *Sputnik.*

Senator Hatch made one more attempt to block the clinic — this time by trying the old congressional ploy of asking for a new report when an earlier one reached a disagreeable conclusion. He tried to obliterate the report issued the previous year by the Ethics Advisory Board — the report stating that IVF research was ethically acceptable, which was still languishing on the HEW secretary's desk — by asking his Democratic colleague Senator Kennedy to hold hearings that would reopen the IVF debate. Hatch wanted to assemble a new commission and make the group start from scratch, looking again at the same ethical and political issues the Ethics Advisory Board had considered. As an alternative, Kennedy proposed writing a joint letter with Hatch to the chairman of the panel that had already been put together to succeed the Ethics Advisory Board, the President's Commission for the Study of Ethical Problems in Medicine and Biomedical and Behavioral Research. In their letter, the senators asked commission chairman Morris Abram to consider reviewing the board's report to see whether it was as dispassionate and thorough as it should have been. If Abram agreed, the new report would be just one more in a long line of reports on IVF and fetal research — and, like the others, it would not be binding on the Norfolk clinic, which received no federal funds.

Abram declined the offer to enter into the endlessly recursive process; the board's report was perfectly adequate, he said, and really didn't need to be looked at again.

Despite all the last-minute attempts to stop it, the clinic finally opened on March 1, 1980. The *Washington Post* was one of many newspapers to praise the clinic — even while acknowledging that its existence raised sobering and possibly paralyzing questions about the nature of human life and human relationships. The successor to the nuclear age, it said in an editorial, would be the age of reproductive biology — the next scientific advance to "shake the foundation of our social values." The editorial writers believed, just as the Ethics Advisory Board did, that the use of IVF by infertile married couples raised "few fundamental ethical questions." But they worried that the technologies that could so easily follow — genetic engineering, for instance, or the artificial womb — could be devastating. Without using the familiar phrase "slippery slope," columns like the *Post*'s put people on alert, warning that once science imposed its will on human reproduction, society would have to think hard about how far it was willing to go.

While the newspapers editorialized, the science itself was taking only baby steps. One incremental accomplishment came in the summer of 1980, when the first Australian test tube baby was born. Candice Elizabeth Reed, born on June 23, was the fourth IVF baby in the world, and she weighed a healthy seven pounds, fourteen ounces. Her mother, twenty-six-year-old Linda Reed, had a reproductive history like Doris Del-Zio's. She too had given birth to one child before her fallopian tubes became irreparably damaged.

The world greeted Candice with nearly the same ebullience with which it had hailed Louise Brown's birth. But there was one surprising skeptic among the enthusiasts. Landrum Shettles, so often accused of drawing conclusions beyond what the data strictly allowed, now accused the Australians of doing much the same thing. In a letter to the editor of *American Medical News,* he stated that the Australian baby might have been conceived without any help from science. "If one reviews the calendar," he wrote, "it appears most likely that this patient, . . . with the emotional uplift from the offer of in vitro fertilization and acceptance in July, 1979, conceived with the following ovulation." Linda Reed conceived naturally, according to Shettles's calculations, in August or September 1979. She believed that her baby had been conceived in October through IVF; Shettles thought it was just as likely that she had entered the IVF program "when she was already pregnant." What threw her off, he said, was that she had had some vaginal bleeding in September — as often happens early on in a normal pregnancy — which she interpreted as a menstrual period. The bottom line, according to Shettles, was that Candice Reed's birth might not have been a scientific milestone at all, just a blessed event for a small family in Australia.

Meanwhile, at the Norfolk clinic, the Joneses and their colleagues were confronting one failure after another. By the end of 1980 they had retrieved eggs from forty-one women and had successfully fertilized and transferred them into the uterus twenty-three times. But they had not achieved a single pregnancy.

"It would have been naive to think we would have success within six months," Howard Jones said gamely in October 1980, at the annual conference of the American College of Obstetrics and Gynecology. But privately he was frustrated, impatient, irritated. He and Georgeanna couldn't quite figure out what they were doing wrong.

Charles Dean said *he* knew what they were doing wrong: they were ignoring more promising therapies that might actually deal with their patients' primary problem, infertility. Dean, who had no medical training (he had made his fortune in lumberyards), said the doctors should be using the Estes operation, a procedure introduced in 1928 to correct infertility caused by blocked fallopian tubes. It involved attaching the ovary to the uterus, meaning a woman would drop mature ova directly into the womb instead of into the fallopian tubes, where they might get hung up in scars and blockages. Most surgeons had abandoned the Estes operation by the 1970s, since it was an invasive procedure and never had a success rate high enough even to measure. But Dean still thought it was better than IVF.

"Why the big push for in vitro fertilization," Dean asked in one of his many letters to the editor of the *Virginian-Pilot*, "when those promoting it know better than anyone else that it is the least likely method by which these women can achieve desperately sought pregnancy? The gruesome answer is found in their insatiable appetite for experimentation on early human beings in the embryonic state — in petri dishes, mounted on slides, frozen, and in whatever other state that tickles their scientific curiosity."

More brazenly than just about any other opponent of IVF, Dean invoked this mad-scientist image, a grisly picture of row upon row of tiny babies struggling against the painful prodding of sadistic men in white coats. The would-be mothers, meanwhile, were what Dean called "little more than convenient vehicles by which a cloke [*sic*] of respectability can be given to this ghoulish activity."

Who needs to be blasted as ghouls and sadists? the Joneses wondered in their darkest moments. Why should they be forced to justify their scientific quest to the Charles Deans of the world?

Dean was not the only critic. Once someone mailed to the Norfolk clinic a grainy photo of Louise Brown with the scrawled words, "She has no soul." Then a fire, whose origin was never really explained, erupted in the office the Joneses shared with Mason Andrews, destroying records and envelopes that contained donations. And Australia's leading right-to-life activist — as if she didn't have her hands full fighting off the burgeoning IVF enterprise in Melbourne — went all the way to Norfolk specifically to denounce the Joneses as "mad monsters [who] really should be locked up."

At times during that difficult year the right-to-lifers' venom stung so badly that the Joneses nearly reconsidered their decision not to retire. Maybe being put out to pasture was exactly what they needed. They wondered occasionally how much more name-calling and political posing the two of them, both approaching seventy, were willing to put up with.

From New York, Raymond Vande Wiele watched the developments in Norfolk with interest. In the two years since the Del-Zio trial, Vande Wiele's position on IVF had softened, just as those of so many other Americans had. The handful of births after Louise Brown's helped convince him that test tube babies were essentially like any other babies. Encounters with colleagues at international meetings helped convince him that infertility would continue to be a pressing problem, especially as members of the baby boom generation reached the age for, and then delayed, childbearing. He was coming to believe that the increase in infertility would color not only the clinical practice of obstetrics and gynecology but also its business environment. Test tube babies, Vande Wiele was starting to realize, were the wave of his professional future.

Vande Wiele sent his Columbia colleague Georgianna Jagiello to Norfolk to visit the Joneses, who were old friends of hers — they seemed to be old friends of just about everybody — and to study their operation. She brought back a full report to Vande Wiele about how the clinic worked, right down to the temperature for incubating the fertilized eggs and the timing of implantation back into the mothers. Vande Wiele listened carefully to Jagiello's report and nodded in that thoughtful way of his. He was making some plans of his own.

At around the same time in 1980, an experiment at Yale was written up in the *Proceedings of the National Academy of Sciences.* Two molecular biologists, Frank Ruddle and Jon Gordon, had successfully added foreign DNA to a one-celled mouse embryo, injecting it directly into the pronucleus, where it became incorporated into the mouse's own genome. The result was a chimera, a mouse that had a little bit of another species' DNA — in this case, a thymidine kinase gene from a human herpes simplex virus — in every one of its cells. Still basically mouse, the chimera had an extra something that made it slightly virus, too. It was, to use the word that Ruddle invented, "transgenic."

The first transgenic mouse showed scientists that they could insert foreign or new and improved genes into the early embryos of other mammals — including humans. For the first time the idea of "designer babies" appeared on the horizon, babies with beefed-up genomes not only in their body cells, but in their sex cells, too — meaning they could pass on their altered genes to subsequent generations. With this small step, Ruddle and Gordon brought the specter of genetic engineering — the insertion of a supposedly "better" sliver of DNA into an emerging human just a few cells in size — out of science fiction and into the realm of possibility.

If the report on the transgenic mouse had been published in a journal with a wider readership — *Nature,* for example, or *Science,* or the *New England Journal of Medicine,* all of which are trolled every week by science writers looking for article ideas — it probably would have generated more debate. But it caused hardly a ripple. Jon Gordon didn't even think he and Ruddle had done anything all that significant: "We just wanted to know, could an animal that had been through thirty million centuries of evolution accept DNA [to which] it had no biological relation and still develop?"

Neither the investigators nor the public talked about the implications of the experiment; who worried about the ethics of manipulating a mouse? But the ethics were important, since the transgenic mouse represented that first step down the slippery slope that so many people had been worrying about. Perhaps oddly, the Yale announcement came just as the IVF experimenters a few states away were closing in on a victory of their own — a victory that would bring us to the brink of the slippery slope, poised for the descent.

19

Opening Pandora's Box

> Science moves with the spirit of an adventure characterized both by youthful arrogance and by the belief that the truth, once found, would be simple as well as pretty.
>
> — James Watson, *The Double Helix* (1968)

Twenty-three, twenty-six, thirty attempts at implantation; twenty-three, twenty-six, thirty extravagant failures. Howard and Georgeanna Jones couldn't figure out why their IVF attempts were going so badly in Norfolk when elsewhere in the world the procedure thrived. How could American ingenuity fail where British and Australian perseverance had succeeded?

The Australian IVF practitioners delighted in sharing just about everything they knew with the would-be baby makers who undertook pilgrimages to Melbourne in the early 1980s, as the city developed a reputation as a hub of high-tech reproductive technology. Melbourne, Australia's second-largest city, was also coming to be known as a capital of culture and good living, with some of the largest shopping arcades and food halls in the country. It was a city of orderly grids and European-style gardens, its citizens likely to be found strolling, sitting in cafés, and checking out the seaside vistas. An old joke told by the residents of Sydney, six hundred miles to the north, typified the friendly rivalry between the two cities: "The only good thing to come out of Melbourne is the Hume Highway." But the joke, like the road, worked both ways: the highway was a route out of Sydney just as much as a route out of Melbourne.

At Monash University, Melbourne's premier institution of higher learning, the medical school was responsible for the operation of several hospitals, including the Queen Victoria Medical Centre. There Australia's first test tube baby was born, as well as its second and third, and the world's first test tube twins. As with many other kinds of pioneering, there was a certain euphoric quality to the IVF enterprise in those days. "There was tremendous excitement when the IVF commenced," said Judy Wood, who at the time was married to Carl Wood, the head of the IVF clinic at the Royal Victoria Hospital. For just about every other medical and technological breakthrough, Australians had learned to go overseas. "Here it was happening right in Melbourne, Australia, and I think that was a really big thrill for everyone."

Members of the IVF Mafia traveled to Melbourne as if to Mecca, hoping to figure out what they could learn — and what they could copy. The odd part was that while investigators from halfway around the world were greeted expansively and handed as much information as they could manage, fellow Australians working at a second IVF clinic less than two miles from the Queen Victoria were completely shunned. Scientists at the two clinics barely spoke to one another.

The Joneses were doing things pretty much the same way the Australians were — with a few important refinements. The most significant was that the Joneses implanted the pre-embryos at a much earlier stage. While the Australians waited seventy-two hours before transfer, by which time the fertilized egg had reached roughly the eight-cell stage, the Norfolk team waited just thirty-six hours, implanting pre-embryos with only two cells. Their timing was based on the results of animal studies showing that cell division is significantly slower in vitro than in the body. Maybe, they reasoned, their mistake had been in waiting too long, until the eight-cell stage at which the embryo usually implants in nature. Better to implant it earlier, maybe even too early, than to run the risk of implanting it too late, when the uterus is no longer receptive. "The embryo will wait for the uterus," Howard Jones liked to say, "but the uterus won't wait for the embryo."

The Norfolk group also used larger-gauge needles to remove the eggs than the Australians did, and had a different method for determining the best time to remove a mature egg. But with all these supposed improvements, the Norfolk team, in their first full year of trying, still encountered nothing but failure.

The Joneses talked endlessly about what they might be doing wrong. "We really don't know if we are dealing with the inefficiencies of in vitro systems," said Howard Jones, "or of human reproduction." Twice, out of sheer frustration, they tried for an "in vivo" fertilization, taking an egg from a woman's ovary and, bypassing her blocked tubes, inserting it directly into her womb, unfertilized, shortly after the woman and her husband had had intercourse. The theory was that the uterus would at that moment be flooded with sperm, all of them eager to penetrate the receptive egg. But in vivo fertilization didn't work for the Joneses any better than in vitro had.

And then, after months of failure and frustration, they stopped talking about it, stopped even thinking about it. That's how breakthroughs often happen, during the calm period when scientists have all but given up, when they've finally managed to empty their minds — or believe they have — until they somehow find, in the interstices, inspiration bubbling through.

At the end of 1980, Howard and Georgeanna Jones left Norfolk to spend the Christmas holiday with their children and grandchildren. Ostensibly they were on vacation. But like many of the best scientists, the Joneses were always working at some level. Freed from the daily chores of running the clinic, away from patients whose disappointed faces, voices, and very bearings seemed sadder and more demoralizing with each passing month, Georgeanna Jones had an epiphany: they must prescribe fertility drugs. They had been avoiding it because of the problems the British and Australian teams had had with the drugs' interfering with implantation, but Georgeanna thought she might be able to avoid this complication. She was an expert in endocrinology, had a long history of success with fertility drugs, and yes, she was going to put her IVF patients on hormones.

The Joneses started to prescribe Pergonal, a powerful fertility drug, in the early part of 1981. The drug regimen was harsh: the three shots a day caused mood swings and irritability and occasionally led to "masculinizing" side effects, including increased hair growth and a deepening of the voice. One of the first patients they put on Pergonal was Judy Carr, a schoolteacher who lived in a suburb of Boston. IVF was against the law in Massachusetts — a liberal state politically, but one with a strong Catholic constituency that kept it from embracing most new developments involving reproduction. It was the last state, in 1972, to legalize the distribution of birth control devices, and in 1975 — two years after *Roe v. Wade* — a Boston obstetrician was convicted of manslaughter for killing a fetus during an abortion. Because of the restrictive atmosphere in Massachusetts, Judy and Roger Carr, who lived less than an hour away from some of the best medical facilities in the country, went all the way to Norfolk to try to make a baby.

Judy, who taught fifth grade, and Roger, an engineer, had met as students at the University of Maine. She was petite and girlish-looking, with short black hair, an animated face, and a thick New England accent. Surrounded all day by other women's children, she was tormented by the feeling that she was being cheated of the large family she and Roger had always wanted. "We both loved children and couldn't imagine going through life without them," she said. They had been devastated by her three ectopic pregnancies — pregnancies that start to develop in the fallopian tube instead of the uterus. Eventually the tubes can burst, ending the pregnancy and endangering the mother's life, so ectopic pregnancies are always aborted, often at the cost of one or both fallopian tubes. After three ectopic pregnancies, Judy was left with no fallopian tubes at all.

When she heard about the IVF clinic opening in Norfolk, Judy didn't think twice about the distance. She and Roger were committed Christians, but they saw nothing wrong with taking this step to have the family they dreamed of. Nor did she gripe about the painful injections of Pergonal, which turned her two-minute commute to school into an hourlong ordeal three times every day. She would start the morning with a twenty-minute drive to the hospital, meet her gynecologist in the emergency room to get a shot, then drive most of the way back to her house to get to Applewild Elementary School, only to repeat the whole procedure during recess and again on her way home in the evening. With her typical high spirits, Judy talked about those assignations as reminiscent of Bob Woodward's meetings with Deep Throat during the Watergate scandal.

After three weeks on Pergonal, suffering from the mood swings they brought on, Judy went down to Norfolk for the next step in the IVF procedure, removal of the eggs. Her eggs matured faster than anyone expected, in just one day, and an apologetic nurse called Judy at her hotel and said, "How soon can your husband get here?" Roger flew down to provide his sperm, and the pre-embryo grew for a few days. On implantation day, April 17, 1980, the day Judy turned twenty-eight, a smiling group of fertility experts carried the petri dish in to her singing "Happy Birthday."

Judy went back to her Norfolk hotel for another week so she could be monitored, and Roger flew back to his job at General Electric. She spent Easter alone, went to the Norfolk Azalea Festival alone, and talked awkwardly to another couple at the hotel, also patients at the clinic, who had just found out that their implant hadn't worked. Georgeanna Jones stopped by on her way to an international fertility conference in Paris and said to her, "Young lady, there is one thing you need to do. You need to go back home and not have a period."

Georgeanna was to deliver a paper at the conference, and Howard went along, as he often did, for the chance to spend April in Paris with his wife. The topic of the Paris conference was not IVF — Georgeanna had been asked to talk about another of her specialties, a form of menstrual irregularity called luteal phase defect — which was a relief for both of them. Usually at such gatherings, because the

Norfolk clinic's mission and its accreting failures were so widely publicized, the Joneses were asked by well-meaning colleagues how the work was going. Each such question was a stabbing reminder that it wasn't going very well at all.

En route to France the Joneses stopped in England to see their old friend Bob Edwards at Cambridge. They spent a day at Bourn Hall, the infertility clinic that Edwards had opened in a renovated country manor, and did not get back to their hotel until late that night. They dutifully phoned their office to check on Judy Carr's situation back home in Boston. "Well, her temperature is elevated," said medical director Jairo Garcia, "and her period is two days late. She's going for a pregnancy test tomorrow."

Had they been in Norfolk, the Joneses probably would have paced the floor until they heard the test results. But the next day they had to fly to Paris, and they did not arrive at their hotel until early afternoon. Howard, who was familiar with Paris from his time as a visiting scholar at the Collège de France, had booked a room at his favorite spot: the Sofitel Bourbon, a small Left Bank hotel just across the Seine from the Place de la Concorde. When the Joneses got there, they found a message to call the office.

It was only eight in the morning in Norfolk, but one of the secretaries, Linda Lynch, was already at her desk when they phoned. Lynch's voice sounded odd, higher than normal, and her words tumbled out. "I'm not supposed to tell you this," she said, "but the test is positive. Judy Carr is pregnant."

Dinner that night was lavishly Parisian; Georgeanna and Howard celebrated the news with escargots, champagne, and enough of their favorite dishes to add up to a bill of fifty dollars, which seemed a fortune in 1981. At a later dinner, with their friend Jean Cohen, the conversation was a bit more somber. Cohen told them details of the first French IVF pregnancy, which had ended in miscarriage, a tragedy after getting so close to happily ever after. He was sure that it was the press's hounding of the couple that had brought on the miscarriage. The Joneses were determined to keep that from happening to Judy Carr.

When they returned from France a few days later, they were greeted at the Norfolk airport by Jairo Garcia and Mason Andrews, who rushed them back to the clinic to arrange a press conference. The Joneses were torn by competing desires. They wanted to protect the Carrs from publicity, but they also wanted publicity for themselves, the clinic, and the medical school. How could they assure secrecy and discretion — especially as they stood there announcing, for publication, the fact that an unnamed woman was on track to have America's first test tube baby? "All we can do is ask your indulgence," Howard Jones said to the journalists who crammed the clinic's tiny conference room. "We know you don't want to be party to an abortion."

After the Ethics Advisory Board sent its formal report to Secretary Joseph Califano of HEW on March 16, 1979, it sat, and sat, and sat, and sat some more. The science was proceeding anyway, financed with private money and beyond the reach of any formal control, while various politicians tried to wish it all away for fear of alienating their antiabortion constituents. The pro-IVF report languishing on the secretary's desk was of no help to the Joneses when they were trying to establish their credibility, nor to Pierre Soupart when he was trying to get funding for his IVF research. HEW became the Department of Health and Human Services (HHS), Jimmy Carter lost the presidency to Ronald Reagan, Califano left, Richard Schweiker came in as secretary — and still nothing happened.

Except for two things. In May 1981, after a year and a half of failed attempts, the Joneses had their first successful IVF pregnancy. And in June 1981, Pierre Soupart, who two years earlier had become the first American scientist to get federal approval for human embryo research, died at the age of fifty-eight, a victim of lung cancer. He had been waiting for a decision on his grant application since 1973.

Australia's third test tube baby, bringing the world tally to seven, was born in the middle of Judy Carr's pregnancy, in the summer of 1981. "People expected her to have a glass bottom," said the delighted father, John Polson. "They expected her to have a glass navel. They would come bearing gifts in test tubes. I mean, it was really an exercise in kitsch." The Polsons named their baby Carla, in honor of the director of the IVF clinic at Monash University, Carl Wood.

The normality of all these test tube babies might have brought the tiniest twinge of disappointment to those looking for excuses to ban IVF or to those who thought that witnessing the birth of an abnormal test tube baby would be something of a thrill, a bit like watching a circus accident on the high wire. But most people felt great relief to see that Carla Polson was just like Louise Brown, just like Candice Reed, just like the other IVF babies: a perfectly ordinary newborn, with ten fingers and ten toes and all her chromosomes apparently intact.

As normal as the first few IVF births had been, however, Howard Jones still worried that the American baby would be the exception. He told no one but Georgeanna and Mason Andrews, Judy's obstetrician, the most terrible of his thoughts: that something might be wrong with the Carrs' baby. The baby's head seemed to be abnormally small, which can be an indicator of a wide range of birth defects. Andrews took frequent ultrasounds of the skull to measure its diameter. Each reading showed that the head was well below the lower limits of normal. Andrews ordered more ultrasounds, and he and the Joneses prayed.

The Joneses wanted to keep their concern a secret from the Carrs, having come to think of the outgoing, spirited Judy almost as a second daughter. But they wanted, too, to keep it a secret from reporters, who would circle around the possibility of a damaged test tube baby like seagulls around a heap of Burger King containers. No one, least of all Howard and Georgeanna Jones, wanted to be involved in the creation of baby monsters. If things went as badly as the Joneses feared, this could be the end of test tube baby making in the United States.

When Judy reached the end of her pregnancy, there was nothing to be done but get the baby out, even though the head was still small. Andrews did take one precaution, however. He planned to deliver the baby by cesarean section; he was taking no chances on a vaginal delivery, with its perilous trip down the birth canal. On the morning of December 28, as Andrews got ready to operate, Howard Jones had the somber task of writing a press release, just in case. Knowing that if the news was bad he would be unable to find the proper words when called before the cameras, Jones wrote it all down: the tragic news of the baby's abnormality, the specifics of the abnormality (he left this part blank, to be filled in when the specifics revealed themselves), his "profound disappointment" about this birth defect, a plea from the parents for privacy and understanding. Then he put the piece of paper in his pocket and prepared to join the dozen or so people assembled in the delivery room. Mason Andrews was there to do the surgery, Georgeanna Jones to stand nearby, Roger Carr to stroke his wife's damp forehead. Also attending the birth were a neonatologist, an anesthesiologist, several nurses, medical students, departmental secretaries, three public TV cameramen — enough of a crowd to make everybody feel that this was truly a historic moment. And while the observers were all atwitter with merriment, they were pretty uncomfortable, too; the thermostat was set at 94 degrees to ease the baby's transition into the world.

The "profound disappointment" press release never came out of Jones's pocket. Instead, after witnessing the birth at 7:46 A.M. on December 28, 1981 — two days before his own seventy-first birthday — and giving a quick cuddle to the beautiful baby with the perfectly normal-looking head, he walked out to the conference room where reporters were waiting. He began with three words, spoken in his deep, soothing, hot-molasses voice. "It's a girl."

The birth of Elizabeth Jordan Carr was met with a kind of public euphoria. Not only were the Carrs delighted; it seemed the whole country was in love with baby Elizabeth. She had a full head of black hair and a perfect little heart-shaped face, and she was, as her smitten neonatologist kept saying at the press conference announcing her debut, "a good baby, a wonderful baby, the kind of baby who would be responsive to you and get you to do things for her." It was clear that her unorthodox start in life, the fact that her parents' gametes had combined in a petri dish instead of in a fallopian tube, hadn't made much difference in the kind of newborn she became.

Elizabeth was the world's fifteenth test tube baby, and although she was the first born in America, she was not the first American. That distinction belonged to Samantha Steel, born on October 2, 1981, to American parents who were treated in England. An estimated one hundred women worldwide, including five in the United States, were thought to be pregnant by IVF at the time Elizabeth was born. "At least five American clinics are treating infertile women in this way, but so far with meager success," wrote Walter Sullivan in the *New York Times* the day after Elizabeth's birth. "Nevertheless, reliability seems to be improving rapidly. The time when it becomes standard treatment for several causes of infertility may be drawing near."

Elizabeth's arrival didn't deter IVF opponents, who were still looking for ways to stop the practice altogether. Three days later, on New Year's Eve, the local Norfolk paper ran an editorial accusing the clinic staff of engaging in a kind of eugenics. For the sake of the clinic's reputation, wrote the editor of the *Virginian-Pilot,* the Joneses insisted that all their patients undergo amniocentesis and agree in advance to abort any pregnancy in which the fetus was likely to be born damaged — no matter what the parents might want.

Those requirements were not, in fact, clinic policy. The truth was, the Joneses were loath to let their patients undergo amniocentesis even when the women wanted to, because in about one percent of cases, the procedure could bring on a miscarriage, and these pregnancies were so hard-won that to lose even one in this way seemed especially cruel. As for requiring abortions of fetuses known in advance to be damaged, Georgeanna Jones had long before gone on record as saying that although she was pro-choice, her personal beliefs made her opposed to abortion. "Everybody is anti-abortion," she said, "but I'm certainly pro-choice because there are times when [abortion] is the right choice."

The *Virginian-Pilot* editorial was the last straw for Georgeanna. She was fed up with having her professional reputation sullied; she hated being turned into a villain, especially with accusations that she was a wanton abortionist. In early May 1982, four months after the editorial appeared, she filed a $5.5 million lawsuit against the newspaper's editor and its publisher, Landmark Communications, one of the richest companies in Norfolk. The case was eventually settled out of court for an undisclosed sum, enough money to keep scientists at the clinic busy with research projects for years to come.

The Norfolk Chamber of Commerce named Georgeanna Jones their Woman of the Year for 1982. The awards luncheon, held on a Friday afternoon in May, featured a fashion show and a local pop singer's rendition of the Helen Reddy hit "I Am Woman." Because of prior commitments in Germany and Spain, Jones could not attend the luncheon herself, but she sent an associate, Lucinda Veeck, to read a statement on her behalf. In it Jones played down the glamorous nature of her job — press conferences, international travel, magazine profiles, awards — and played up the traditional role she had always relished. "There is no more demanding or responsible career than caring for one's children and the homes we love," she wrote. "Ladies, we who enjoy our out-of-home careers and have contributions to make in other areas must see that the career of homemaking and child-rearing is equally recognized."

The speech was a lovely variation on the paradoxes of the early eighties. Here was Georgeanna Jones, who had excelled in a difficult medical specialty dominated by men, who had made her mark helping career women overcome infertility created partly by their pursuit of those very careers, who could not attend her own award luncheon because she was off on a European lecture tour. Yet what she chose to talk about was not women's liberation or the Equal Rights Amendment or the concept of equal pay for equal work. Instead, she talked about the peculiarly sweet, feminine, domestic, homebound, private, quiet joys of bearing and nurturing your own blossoming babies — no matter how they came into the world.

In September 1982, less than nine months after Elizabeth Carr was born, the *Washington Post* published a state-of-the-science wrap-up article with the latest tally: fifty-four children born in England in Edwards and Steptoe's clinic, with another thirty-seven pregnancies under way; thirty-three born at Monash University in Australia, with another twenty-three pregnant. By the spring of 1983, projections were that one hundred fifty test tube babies would have been born, "making yesterday's miracle commonplace."

20

Tables Turned

> In the blastocyst, even in the zygote, we face a mysterious and awesome power, a power governed by an immanent plan that may produce an indisputably and fully human being.
>
> — Leon Kass, "'Making Babies' Revisited" (1979)

In the greater scheme of things, ten years is a tiny blip of time, a matter sometimes of a single digit that schoolchildren confuse when memorizing dates in history books. Even from the vantage point of less than a generation, we sometimes have trouble keeping decades straight. How is a Baby Boomer born in 1949 different from a Baby Boomer born in 1959? Which ten-year span was called the "Me Decade" and which the "We Decade"? What does a Generation X-er have in common with someone from Generation Y?

But as one is living through a decade, it can seem to last forever. It took Edwards and Steptoe ten years from their first successful laboratory fertilization of a human egg until the birth of an actual IVF baby. It took a total of ten years, in two widely spaced bursts of productivity, for Michelangelo to paint the ceiling of the Sistine Chapel. And it took ten years, and the birth of a few dozen healthy test tube babies, for Raymond Vande Wiele to completely change his mind about in vitro fertilization.

By 1983, ten years had passed since John and Doris Del-Zio arrived with her suitcases, full of hope and trepidation, on the steps of New York Hospital. During this single decade, this blink of an eye in objective time, a silent revolution had taken place in society's view of children like Louise Brown, who by now was almost ready for kindergarten. Test tube babies had gone from being a risky and bizarre idea to being ordinary little everyday miracles — so ordinary, in fact, that Lesley and John Brown had gone back for another IVF and had given Louise a little test tube baby sister named Natalie.

And now Columbia University, the site of the controversy over the Del-Zio IVF attempt, was about to open an in vitro fertilization clinic of its own. Although there were by now six such clinics around the country — at Eastern Virginia Medical School, the University of Southern California, the University of Texas Medical Center, the University of Pennsylvania, Vanderbilt, and Yale — this would be the first IVF facility in New York. It was to be located at Sloane Hospital for Women, a Columbia affiliate. And it was to be headed, strangely enough, by Raymond Vande Wiele.

In an interview with a reporter from the *New York Times* announcing the clinic's opening, Vande Wiele didn't mention his jagged personal history with IVF. He didn't say a word about the Del-Zios or their lawsuit against him; he never mentioned the name Landrum Shettles. He was, as usual, reserved and distant, saying only that he and his co-director, Georgianna Jagiello, had selected twenty-five potential patients out of an applicant pool six times that number. Attempts at laboratory fertilization were to begin within the month, he told the reporter, Judy Klemesrud. But these attempts were no small matter: each one could drag on for "eighteen, nineteen, or twenty days," Vande Wiele said, at a cost of some five thousand dollars per cycle. Most of the expense was not covered by insurance, and the psychological burden was, of course, something that only the would-be parents could bear. Despite these obstacles, however, infertile couples were clamoring for IVF — as evidenced by the "significant number of pregnancies" around the world that Vande Wiele said would soon lead to a doubling of the international test tube tally.

Characteristically, in his *Times* interview Vande Wiele said nothing of the ethical quandaries he and his peers had wrestled with over how far they should go in manipulating life; nothing of the government's continued refusal to support IVF research, even after some fifty healthy IVF babies had been born around the world; nothing of the two dozen states that had passed laws limiting or prohibiting research involving fetuses, about half of which banned research on IVF and embryos, too; nothing of any conflict between his Catholic upbringing and his desire to give patients the babies they so clearly craved. He simply told Klemesrud, in his dry and understated way, "This is a formidable procedure."

Nonetheless, he was willing to go out on a limb — an indication of the high stakes of this competitive business, with Columbia edging out Cornell and its affiliate, New York Hospital, by only a matter of months in the race to open the city's first IVF clinic. Vande Wiele used the opportunity to put in a good word for his own enterprise and to differentiate it in the public mind from any New York City clinic still to come. In vitro fertilization, he said, held out great promise for the nation's childless couples, and he expected a success rate better than that of Howard and Georgeanna Jones in Norfolk, or of Edwards and Steptoe in Oldham, or of Carl Wood and his colleagues in Melbourne. He expected the Columbia team to achieve one pregnancy for every five attempts.

And he made one bold prediction that was an even greater surprise, especially considering how recalcitrant he had been about getting this far in the first place. In another ten years, Vande Wiele said, scientists might be able to correct genetic defects in an IVF pre-embryo before implantation, while it was still in culture outside the womb. At the time, such a prospect was particularly frightening. It implied the most profound manipulation of the human gene pool, raising the specters of eugenics, selective breeding, and the elimination of supposedly "bad" genes whose potential benefits no one fully understood. Yet in early 1983, here was Raymond Vande Wiele predicting that the in vitro correction of genetic defects would happen within a decade — and calling this controversial accomplishment "the ultimate goal" of IVF.

It was as though the slippery slope had been given an extra coating of motor oil. Genetic manipulation as the "ultimate goal"? Altering the human germ line — the eggs and sperm — in a way that would have profound and permanent effects on all the descendants of the unborn baby? Allowing ordinary, human scientists to become, in essence, arbiters of evolution, taking upon themselves the godlike role of deciding which genes should persist and which should die out? How strange for Vande Wiele to toss off this idea as though it were an inevitability. Once a symbol of scientific caution and procedural propriety, he now seemed to be saying that if you want a perfect designer baby, as long as you come to the Columbia clinic, anything goes.

Vande Wiele never had a real chance to reinvent himself as an IVF *consigliere.* The clinic opened in 1983 with a waiting list of 150 anxious couples, and Vande Wiele and Jagiello worked well together. But it was a short-lived partnership. In August, seven months after the clinic opened, Vande Wiele went to Morocco to assist in the queen's medical care, and from there traveled to Seattle for a medical conference. Friends who accompanied him to dinner in Seattle, among them Luigi Mastroianni of the University of Pennsylvania, thought that Vande Wiele looked exhausted and drawn and did not seem to take his usual hearty pleasure in the excellent food and wine. Vande Wiele admitted to feeling a bit tired, nothing more.

The next day, after returning to New York, he knew something was wrong, so he made an appointment to see his internist the following Monday. On Saturday he went to a scientific slide session with an old friend, Michel Ferin, whose father had been a teacher of Vande Wiele's in Belgium and who now worked in Vande Wiele's department at Columbia. Ferin thought his friend looked tired, too, but he didn't say anything. Later that evening Vande Wiele took his wife, Betty, to see a Woody Allen movie.

In the wee hours of the next morning, Sunday, August 14, 1983, Raymond Vande Wiele woke up with severe chest pains. His wife called an ambulance, and he was taken to Englewood Hospital, where he died. He was sixty years old.

21

From Monstrous to Mundane

Man gets used to everything — the beast.

— Fyodor Dostoevsky, *Crime and Punishment* (1866)

Since the birth of Louise Brown in 1978, in vitro fertilization has become almost routine. In the twenty-first century it may be one of the first steps on the fertility merry-go-round rather than the last. Many medical insurance plans cover it, and people going through the treatments talk about it. IVF is no longer monstrous; today it is almost mundane.

Of all the forms of assisted reproductive technology (ART), defined by the Centers for Disease Control (CDC) as any procedure in which the egg and sperm are handled outside the body, in vitro fertilization is the most common. In 1998 nearly 82,000 IVF procedures were done in the United States, giving rise to more than 28,000 test tube babies that year (impressive numbers — but they also mean two out of every three procedures failed). Since the mid-nineties, the number of IVF clinics nationwide has grown by more than 36 percent, from 330 in 1996 to an estimated 450 today.

Who could have anticipated this explosion of demand for IVF when it all began? In the 1970s in vitro fertilization was often portrayed as the bluntest possible intrusion into something intimate, inviolate. Even those who thought it was a good idea, who believed in using technology to impregnate women with blocked tubes, could see that it involved mucking about in the very stuff of life. Scientists might have thought it was perfectly acceptable to manipulate human tissue in this way and that doing so carried few genetic or medical risks. But at some level many of them, even the most gung-ho, worried that their research was crossing some elemental line.

Today scientists doing in vitro fertilization don't think much about elemental lines. When a fertility expert at an IVF clinic wiggles a probe to retrieve an egg or squints into a microscope to watch the "dance of love" play out in a petri dish, she probably is not asking herself questions about when life begins or who should play God. She is just doing her job, trying to get the conditions right, thinking about the culture medium and the incubator temperature and the next step in the procedure. Her ultimate goal is to make a baby, and after the births of hundreds of thousands of babies made in just this way, questions about the rights and wrongs of such manipulations, questions about whether the ends justify the means, are virtually never asked.

So many new variations now exist that plain old white-bread IVF — the kind used on the Del-Zios and the Browns — seems almost dull. The variations have charming names and acronyms: GIFT (gamete intrafallopian transfer), ZIFT (zygote intrafallopian transfer), zona drilling, assisted hatching. GIFT and ZIFT can both be thought of as IVF lite. Whereas basic IVF involves mixing the gametes in a petri dish and culturing them after the egg has been fertilized, GIFT just puts the gametes together in the fallopian tube, so that fertilization can take place in the body. ZIFT, instead of transferring the zygote directly into the uterus, places it in the fallopian tube, which gives the zygote more time in the body before it implants in the womb. And unlike basic IVF, in which the sperm makes its way into the egg on its own in the petri dish, zona drilling and assisted hatching give the sperm a little edge, through microscopic manipulations of the egg's protective coat to allow even a slow-moving sperm to penetrate.

IVF may have become routine, but it still isn't easy. Pregnancy rates from IVF remain low — about 24 percent on average in the United States, according to the CDC — and the birth rates are even lower. These rates are not much different from Mother Nature's, but they seem worse because the attempt itself affords no pleasure — only inconvenience, worry, discomfort, and expense. A woman doesn't usually count the number of times she makes love when she's trying to conceive — all but the most attentive think in terms of months, not sexual encounters — so she doesn't really know the denominator of any fraction involving her chances of getting pregnant any one time she tries. But with IVF she counts, and waits, and prays, and gets a bill for each individual attempt.

Nothing in the process of artificial conception even vaguely resembles lovemaking. And every step carries risk. The woman is stoked with hormones to stimulate the maturation of multiple eggs, and these hormones may increase her risk of cancer or other problems later in life. There's also a more immediate hazard: some 1 to 10 percent of women taking fertility drugs (the exact rate is still being debated) develop a complication known as ovarian hyperstimulation syndrome. In these women, from whom as many as twenty or thirty eggs can be harvested in a single cycle, the ovaries become so overloaded that they may actually burst. As the surgeon pulls out egg after egg after egg, he may be glad to have so many chances to get a good zygote, but he also knows that something is probably wrong. Most fertility specialists consider ovarian hyperstimulation syndrome to be a failure of IVF.

A failure, too, is the birth of more than one baby at a time. Not to the parents, necessarily; they often think that being pregnant with twins is like getting two for the price of one. With the news that more than one implanted embryo has "taken," many couples resonate to the idea of an instant family, a chance to have all the children they've dreamed of without going through the agony and expense of additional IVF cycles. How wonderful it will be, they think, to gather up their multiple blessings, to revel in the noise and the soft, sweet smells of those delicious babies. Forgotten for the moment are the logistical headaches and emotional strains that will surely come from juggling the feeding, diapering, and cuddling of more than one newborn at a time. Also forgotten sometimes is the more significant problem: that a multiple pregnancy is a high-risk pregnancy. Twins are twice as likely as singletons to die within the first year of life; triplets are thirteen times as likely. Multiples have higher rates of prematurity, low birth weight, and cesarean section, all of which are significant causes of birth defects and infant mortality. And they are more likely than singletons to suffer from blindness, cerebral palsy, mental retardation, and other significant anomalies.

Throughout the 1980s, test tube broods kept getting bigger, accumulating a startling list of "firsts." The world's first IVF quadruplets were born on January 6, 1984, in Melbourne. The babies' mother, a thirty-one-year-old woman known as Mrs. Muir, entered the Royal Women's Hospital shortly before Christmas and was delivered by cesarean section two weeks later. The babies, all boys, were six weeks premature, and weighed between 3.9 pounds and 4.6 pounds. Their parents named them Sam, Ben, Christopher, and Brett — and, they admitted, they honestly couldn't tell which one was which.

Soon test tube quads were not newsworthy to anyone but the dumbfounded parents and the legions of friends and neighbors recruited to help out with their care. The real news was test tube quintuplets. The world's first, also all boys, were born in London in 1986; the first in America, in 1988, just outside Detroit; the first in Canada, one month later, in Toronto.

Multiple births have been a hazard of IVF from the beginning, when doctors routinely returned six or eight or even ten embryos to the womb in hopes that at least one would implant and develop into a baby. This practice was a way to bolster the success rates by which most clinics were judged. The more embryos in the uterus for any one cycle, the more likely it was that the cycle would result in a pregnancy — giving the clinic's statistics, calculated as pregnancies per cycle, a nice little surge. But with so many embryos in the uterus, not only did the odds increase that one would implant; the odds also increased that more than one would.

Even today IVF practitioners in the United States put back more embryos than most professional groups think they should. According to surveys, they transfer three or more embryos at least 80 percent of the time; nearly half the time they transfer more than four. As a result, 56 percent of IVF births in the United States are twins, triplets, or more. In Europe, on the other hand, doctors transfer three or more embryos in only half of all IVF cycles, and four or more in just 9 percent, and the rate of multiple births is correspondingly lower than in the United States. Only 24 percent of European IVF babies born are twins, and only 2 percent are triplets, quads, or more — about half the rate of multiples seen in the United States.

Why are American clinics still transferring so many embryos at once? Because they are still being judged by their statistics — statistics that may determine whether patients will choose one clinic over a competitor's. Howard Jones has gone so far as to suggest a change in the law that currently requires the CDC to publicize clinics' pregnancy rates. The law was passed in 1992 to offer patients an objective, government-sanctioned set of data with which to differentiate among clinics of varying quality. Jones was one of the early proponents of such a law. But he came to believe that it might provide the main motivation for transferring too many embryos in pursuit of those better rates. Changing the law now, according to Jones, might take away that motivation and might thereby reduce the disturbing number of IVF multiple births.

For many, the darkest side of IVF in the new century has changed. The concern today, at least for most people, is not that mankind is treading on God's turf, nor that the bastions of civilization will crumble, nor that we'll soon be tripping down a slippery slope toward gross indecencies. These scenarios are connected in the public imagination with human cloning rather than IVF. The biggest concern about IVF now is medical rather than ethical or social or political: the concern that it might be causing birth defects.

There have been inklings of long-term problems for years. Small-scale studies in the late 1980s and early 1990s found that the rate of miscarriage for IVF pregnancies was two times higher than normal; of stillbirths and neonatal deaths, three times higher; and of ectopic pregnancies, five times higher. An IVF baby, some studies showed, was five times more likely to be born with spina bifida, a birth defect involving the brain and spinal cord, and six times more likely to be born with a particular heart defect known as transposition, in which the chambers of the heart are on the wrong side. And in the refinement of IVF known as GIFT, the rate of stillbirth and neonatal deaths was especially high — five times higher than normal.

Strangely, these findings had almost no effect on the steady increase in customers for reproductive technology. It was almost as if people were hearing only what they wanted to hear. This happens often in questions relating to public health when scientific studies contradict each other and people are left to draw their own conclusions about, say, estrogen replacement therapy, salt in the diet, margarine versus butter, or mammography. With IVF there were plenty of positive studies to counterbalance the worrisome ones. Studies from England, Australia, France, and Israel all found that IVF babies ran about a 2 percent risk of birth defects, no higher than in the general population — and for people predisposed to think good thoughts about IVF, these were the ones that held sway.

What a change this was from the public attitude toward in vitro fertilization in its early days. When the technology was new and under assault from all sides, any single misstep, however minor, would have been enough to quash it altogether. If Louise Brown or Candice Reed or Elizabeth Carr had had the slightest thing wrong with her, in vitro fertilization would have been blamed, and the technique would have suffered irreparable damage. But as IVF became routine, the thinking shifted. The risks, if any, were vague, hard to document, and difficult to prove, while the benefits were there in front of everybody's nose. All those luscious babies assembled on clinic lawns for anniversary picnics; all those newborns in the arms of parents who had thought they never would hold a child of their own; all those laughing, running, living, breathing children who would not have existed if not for the miracles of modern medicine — wouldn't it have been churlish to dig beneath the surface for evidence that these perfectly normal-looking kids had something wrong with them?

By 2002, however, it was impossible to ignore IVF's dark side any longer. In early March two unrelated studies were published together in the *New England Journal of Medicine.* They were among the first to take into systematic account the higher rates of multiple pregnancies with IVF in assessing its overall risk — a complicating factor, because multiples are themselves at higher risk of birth defects. One study compared the birth weights of more than forty-two thousand IVF babies born in the United States in 1996 and 1997 with those of more than three million naturally conceived babies. Excluding premies and multiples, the test tube babies were still two and a half times more likely to have a low birth weight, defined as less than 2,500 grams, or about five and a half pounds. The other study looked at more than five thousand babies born in Australia between 1993 and 1997, 22 percent of them born after IVF. It reached much the same conclusion: the IVF babies were twice as likely as naturally conceived infants to have multiple major birth defects, in particular chromosomal and musculoskeletal abnormalities. The rate of birth defects overall was close to 10 percent among IVF babies — an increase, the Australian researchers speculated, that might be connected to the drugs used to induce ovulation or maintain early pregnancy or to the infertility problems that brought the couple to IVF in the first place, which might have some unknown connection to birth defects. And, they acknowledged, it might be the technique itself that somehow was to blame.

Suddenly it was as if a dam had broken, and a number of reports emerged of higher rates of several kinds of birth defects in the IVF population. In November 2002 scientists at Johns Hopkins and at Washington University of St. Louis found that a rare birth defect known as Beckwith-Wiedemann syndrome was six times more likely to occur in IVF babies than in the general population. The syndrome is a chromosomal abnormality thought to reflect a problem with genetic imprinting — the way a cell recognizes whether its genes come from the mother or the father. Imprinting is an important part of the process through which each cell turns on only the appropriate genes to become whatever type of cell it is supposed to be. One hypothesis is that this process, known as gene expression, is interrupted when gametes are manipulated in the lab.

Beckwith-Wiedemann syndrome is most likely to occur in IVF babies if they were conceived through the variation known as ICSI (pronounced ick-see), the shorthand for intracytoplasmic sperm injection. A hair-thin needle is used to pierce the egg's corona and, guided by a powerful microscope and a steady hand, inject a single sperm directly into the ovum. With this technique, a sperm that is too weak to swim up the fallopian tube or even to penetrate an ovum in a petri dish is shot straight into the egg. To watch ICSI is to be awed by the double brilliance of both science and nature. How staggering is the grace of a sperm performing its magic; how stunning, too, the precision with which a manmade procedure is able to imitate that grace.

But giving a weak sperm this much power — power that it would not have had normally in trying to make its way up the reproductive tract — may bring about problems, not only Beckwith-Wiedemann syndrome, but other, even more serious birth defects. ICSI is used only when a man's sperm is too misshapen or impotent to penetrate an egg on its own, even if it's delivered right to the egg's front door. Some people worry that by enabling severely weakened sperm to do a job that's usually accomplished only by the hardiest, scientists are passing on certain kinds of damage to the next generation. ICSI turns everything about natural selection upside down. If scientists allow for survival of the unfittest in this way, are they increasing the risk of abnormalities? Or, at the least, are they creating babies who will themselves produce abnormal sperm?

In March 2003 scientists at Johns Hopkins noted a sevenfold increase among IVF babies of a severe birth defect known as bladder exstrophy, in which the bladder is inside out and protrudes through the abdomen. Also, the skin on the lower abdomen does not form properly, the pelvic bones are widely spaced, and the genitals may be abnormal. Like Beckwith-Wiedemann syndrome, bladder exstrophy is extremely rare, making it hard to draw conclusions about its association with IVF based on a small sample size.

Other questions are being raised about the long-term risks of embryo freezing, which is often used in IVF. When IVF began, clinics avoided freezing embryos altogether so they wouldn't have to ponder the possibilities raised by this Pandora's box: preimplantation diagnosis, prenatal adoption of embryos, the birth of a baby months or even years after its father's death, the eventual disposal of embryos nobody wanted. Still, despite the early reservations, embryo freezing became popular almost as soon as it was introduced. People might not have wanted to consider its ethical implications, but they did seem to want its convenience. The first frozen embryo birth took place in Australia in 1984, and by 1988 sixty-seven American clinics had cryopreservation programs, nearly ten thousand pre-embryos were in cold storage, and eighty-three babies had been born from previously frozen embryos. Today the number of frozen embryos in storage in the United States alone is estimated at four hundred thousand.

The vast majority of clinics now routinely freeze their patients' extra embryos. Not to offer cryopreservation would be bad business — one more example, like the transfer of too many embryos, of an IVF practice being driven not by scientific considerations but by the bottom line. Patients expect embryo freezing, and if they don't find it at Clinic A they will take their business to Clinic B. They think of their frozen embryos, trapped in suspended animation in fingernail-sized vials, as a kind of insurance policy. Freezing is "an emotional boon to patients," wrote one observer. "It will save patients from having to take hormones that make them edgy, travel to strange cities, and undergo laparoscopies. The couples can relax with the knowledge that they have embryos. . . . Patients are excited about frozen [embryos]. They don't have to be monitored or stimulated each time . . . [They] know they have something in the tank. They are happy about that."

But concerns about cryopreservation have emerged in recent years — concerns once again about medical rather than ethical complications. "Basic functions such as growth, respiration and metabolism are regulated by genes," said Lord Robert Winston, one of Edwards and Steptoe's opponents in the early 1970s (he issued a public apology to them after the birth of Louise Brown). "And if you change the way those genes are expressed — even temporarily — during times of rapid development, such as an embryo, you may well expect to see changes in the way the embryo develops." Winston said his laboratory at Hammersmith Hospital in London had found that freezing embryos affected "the normal activity of vital genes." The long-term implications of such findings are not yet known.

Despite the uncertainties, IVF today seems to have a Teflon reputation. It is still considered a simple, necessary, and basically harmless procedure, a medical intervention with the same sort of risks and benefits as any other intervention. Maybe the IVF patients' passion for babies makes them overlook the possible dangers of the path they've chosen. Or maybe they go into the enterprise with eyes wide open, having carefully considered their options and decided that IVF is worth the risk.

Imagine such a couple sitting at their kitchen table, talking into the night about whether to try in vitro fertilization. They might pull out a pad of paper and carefully list the pros and cons. On the pro side are that a baby will bring great joy to their marriage; that the odds of having a normal baby through IVF greatly outweigh the odds of having a damaged one; that many of the problems uncovered by recent studies, such as low birth weight, could be relatively mild; that adoption is a difficult process. On the con side are the low but real risk of having a baby with a serious problem such as spina bifida or heart defects; the long-term hazards, which are still unknown, of fertility drugs; the strain on the relationship from depersonalizing one of marriage's most intimate acts; the time; the expense; the odds against it even working. Despite all the negatives, the couple might reasonably conclude that the risks are worth it for the sake of the big payoff — a baby of their own.

It remains an open question whether these prospective parents should be allowed to make such choices without restraint. Is there a cost to letting them opt for potentially dangerous treatments? Babies with serious birth defects are a heartache for parents, but they are trouble for the rest of us, too; they siphon off a disproportionate share of the nation's medical resources. Does society, which will bear the brunt of paying for damaged babies, deserve a voice in the decision? What about men with abnormal sperm who opt for ICSI, or women who decide to try IVF even as menopause approaches? Does society have a stake in limiting the reproductive options of these high-risk individuals?

Even if we do have a stake in the outcome, we don't put restrictions on couples who decide to get pregnant without medical help. We don't tell them that the man's weak sperm or the woman's advanced age puts them at higher risk for problems, and that we won't allow them to try to conceive in the privacy of their own homes. Many consider the right to have children a God-given right; others think of it as a right that is protected in the Constitution. To use the mother's age or the father's sperm quality to decide when to abrogate that right smacks of eugenics. Why should it be any different when conception is done in a laboratory? Don't the high-risk couples who go to infertility specialists deserve the same respect as the high-risk couples who try to make their babies in bed? The issue remains, no matter where conception occurs, a matter of whose rights take precedence: society's right to keep its health-care expenditures under control, or the couple's right to the begetting, through whatever means necessary, of a child.

22

Pandora's Clone

> . . . A world of made
> is not a world of born — pity poor flesh
> and trees, poor stars and stones, but never this
> fine specimen of hypermagical
> ultraomnipotence . . .
>
> — E. E. Cummings, "pity this busy monster, manunkind" (1944)

Two days after Christmas 2002, a French-born scientist named Brigitte Boisselier held a press conference in Florida that one journalist described as "surreal." Boisselier, a forty-six-year-old chemist, looked a bit like Zsa Zsa Gabor, if Zsa Zsa had burnished her hair, grown it to her waist, and outlined her lips with eyebrow pencil. She said she was a "bishop" in a religious sect known as the Raëlians, who believed that the first humans were cloned by space aliens 25,000 years ago. Boisselier was also the chief scientist of Clonaid, a subsidiary of Valiant Ventures Ltd., and she had assembled the press in a crowded Holiday Inn meeting room to announce the birth of the world's first human clone.

The clone was a healthy baby girl, said Boisselier, and the Raëlians were calling her Eve. The cloning had taken place, she said, in a country she would not reveal, by a doctor she would not identify, on a woman she would not name. All she would say about the DNA donor was that she was thirty-one, an American, and married to a man who was sterile. Boisselier promised to provide irrefutable evidence that the woman and her baby were genetically identical, but she did not specify how. The Raëlians had four more clone pregnancies under way and about one hundred couples waiting in line to be cloned. Boisselier said that two of the four clone pregnancies had been done, as this one was, because of infertility, and the other two were clones of babies who had died, whom their parents were desperate to replace.

"Absolutely abhorrent, unsafe and ethically questionable," said Robert Lanza of Advanced Cell Technology (ACT), when interviewed on CNN about the alleged cloning. The irony is that the same sentiment had been expressed just a few months earlier about Lanza's own company, when scientists at ACT announced that they had cloned a human embryo for research purposes and allowed it to grow to the six- or eight-cell stage.

"Irresponsible medicine," said University of Wisconsin ethicist Alta Charo, a veteran of several national bioethics commissions, on the same CNN program. "It's a truism in medicine," she said, "that you don't begin working on humans until you have a basis in laboratory and medical science." This same objection had been raised, almost verbatim, by critics of IVF in the 1970s.

Boisselier's unsubstantiated claim was not much different from the cloning claim made by David Rorvik in 1978, or the test tube toddler claim made by Douglas Bevis in 1974, or the lab-cultivated fetus claim made by Daniele Petrucci in 1961. It attracted a great deal of attention for a few weeks, not least because the Raëlians were a colorful sect, with their stories of UFO abductions, their white Star-Trekian outfits, their plans to build a Welcome Center for space aliens in Jerusalem. Their true goal, the Raëlian leader admitted in one of the dozens of newspaper interviews he granted during the publicity blitz, was not just making cloned babies, but "creating scientifically from scratch a completely synthetic human being."

True or not, documented or not, possible or not, Boisselier's announcement of a human clone raised many of the same ethical, moral, religious, and societal questions raised earlier by IVF — questions about whether we as a community dared allow such a step.

"Talk about this baby not like a monster, like some results of something that is disgusting," Boisselier urged at her Florida news conference, asking reporters to tread carefully on the feelings of the woman who had given birth to Eve. She called this woman the baby's "mother," but of course she was not her mother in the traditional sense of someone who had contributed one-half of the baby's genes. And the lack of anything more accurate to call her was just the first indication of how complicated this cloning journey was going to be.

Foes and advocates of cloning speak of "therapeutic cloning" and "reproductive cloning." But this semantic tethering is a result of politics rather than science. Cloning foes want to emphasize that creating human embryos in the lab is morally equivalent whether the goal is research or producing a human clone; advocates want to underscore the potential medical significance of the research itself. But the two types of cloning are in most ways unconnected except in the popular imagination. And their greatest difference lies in what each type might achieve.

Therapeutic cloning is a laboratory technique that propagates cells in a petri dish to meet a particular patient's need for replacement parts. Reproductive cloning uses the same technique, but with the intention of making a baby that is genetically identical to someone else. In therapeutic cloning the embryo is a way station, a developmental phase in the culturing of individual cells. In reproductive cloning the embryo is the whole point.

Therapeutic cloning is tied to embryonic stem cell research; the embryo is created for its stem cells, and it is made through cloning rather than IVF to be sure that its DNA is identical to the donor's DNA for the sake of immune compatibility. The scientist begins by removing the nucleus of a human egg — the nucleus being the place where the genes reside — and replacing it with the nucleus of a skin cell or other body cell. The renucleated egg is stimulated so that it will divide and grow, thus creating a culture of cells that are in effect a laboratory embryo, with every cell having a genome identical to that of the donor of the original nucleus. The cells grow for four or five days until they form a blastocyst, that tiny ball of a few hundred cells wrapped around a liquid center like a bonbon. The fluid inside the blastocyst is made up of stem cells, which are extracted and grown in a new culture. These are the cells of interest here. Stem cells are "pluripotent," meaning they can become any kind of cell as the embryo differentiates into a fetus, with limbs, bones, brain, and organs. They can become any kind of cell in the petri dish, too, and scientists are learning how to encourage stem cells in culture to transform into a particular type. With therapeutic cloning scientists hope someday — and that day is still far off — to be able to take a skin cell from a patient with, say, Alzheimer's disease, grow an embryo in a petri dish using the nucleus of that cell, and transform a few of those embryonic stem cells, each of which has the potential to turn into any cell at all, into brain cells. These healthy brain cells could then be injected into the patient — without any concern that his body will reject the new cells, since they are genetically identical to his own.

Reproductive cloning starts out in exactly the same way as therapeutic cloning: take the nucleus out of an egg cell, replace it with the nucleus of a body cell, stimulate the new egg so it starts to divide and form an embryo. The difference is that after a few days the embryo is taken from the dish and implanted into the uterus of a willing woman, just as occurs with IVF. Eventually a baby clone is born, with exactly the same genome as the donor of the original nucleus.

Almost everyone who has been asked — scientists, politicians, religious leaders, people on the street and in the grocery store — wants human reproductive cloning to be banned. In 1997, shortly after the birth of Dolly the sheep, the first mammal to be cloned from an adult cell, a Time/CNN news poll asked one thousand Americans whether cloning was "against the will of God." Three-quarters of the respondents said yes. Many people object to therapeutic cloning on similar grounds. Those who want to ban it — in most cases people whose religion says that life begins at conception — oppose therapeutic cloning for two reasons: because they see it as creating humans just to kill them for their parts, and because they believe it's only a short slide from therapeutic cloning to having someone secretly slip one of those laboratory clones into a woman's womb.

Syndicated columnist Charles Krauthammer is one of the most forceful proponents of this view, the ultimate in slippery slope arguments. "Banning the production of cloned babies while permitting the production of cloned embryos" is logically inconsistent, he writes, and will lead only to chaos. Like others who oppose all forms of cloning, Krauthammer says it is "inevitable" that one day, without anyone really even noticing it, one of the embryos created for research purposes will ultimately be implanted in a woman's uterus — or, even worse, in an artificial womb in a laboratory. At that point, Krauthammer writes, legislators would be in a real bind: if they have passed a law prohibiting reproductive cloning, they will have made it a crime *not* to abort the clone growing in that woman or that artificial medium. How ironic that the very people opposed to embryo research because it involves the destruction of what they see as nascent human life would be left in a position to force a woman pregnant with a clone to have an abortion.

To Krauthammer, therapeutic cloning would send us down a slippery slope toward not just reproductive cloning but a whole cesspool of murderous behavior. From tearing up blastocysts, according to Krauthammer, it's just a baby step to tearing up a "recognizable human fetus," or maybe even an infant, for spare parts that might benefit the already-born. "Violate the blastocyst today," he writes, and you'll be "violating the fetus or even the infant tomorrow." No matter how much you dress up the cloning of embryos in the laboratory, says Krauthammer, no matter how much you try to justify it by talking about how many degenerative diseases the embryonic stem cells might one day be used to cure, the practice can be reduced to its most vile essence: a cottage industry dedicated to the creation of human embryos that can be used as fodder for what Krauthammer calls "dismemberment for research." His inflammatory word choice must be deliberate. He has to know, since he is a physician as well as a journalist, that a two- or three-day embryo has nothing even vaguely resembling body parts that can be "dismembered" — it's just a collection of cells, a tissue culture, not a thing with shape. The word he chooses may be emotional dynamite, but it's scientific doublespeak.

Language both reflects and shapes the public perception of biomedical truths. That is why partisans in the cloning debate take so much care to use loaded terms like "therapeutic cloning" or "embryo farms." Even the very first step, the cell culture itself, is given different names depending on the speaker's point of view. Technically it is an embryo, which is defined as a conceptus younger than eight weeks postfertilization. But most people think of "embryo" as equivalent to "fetus," a much more mature and differentiated entity. So cloning proponents try to avoid calling it an embryo, choosing instead to call it a pre-embryo, an early embryo, a zygote (though a zygote, strictly speaking, is a fertilized egg cell; as soon as it splits from one cell into two, it stops being a zygote), a morula, a blastocyst. All of these terms have less emotional resonance than "embryo."

Public perception is also shaped by popular culture, by books and movies that are notoriously inaccurate when it comes to science. Poetic license has mangled many people's understanding of cloning, leading them to expect a clone to be identical in every way to its donor, springing full grown like Athena from the head of Zeus. But cloning does not mean xeroxing people so they'll be identical versions of the original — or even a little smudgier than the original, like xeroxes themselves, like the successively more dimwitted versions of Michael Keaton in the 1996 film *Multiplicity.* Nor does it mean breeding whole battalions of interchangeable creatures — soldiers or laborers or prostitutes or whatever else a despotic leader might desire — as described in Aldous Huxley's 1932 novel *Brave New World* or in the bestseller written by Ira Levin more than forty years later, *The Boys From Brazil.*

Despite the popular mythology, the truth is that just because a clone has the same DNA as its genetic donor does not mean it will grow up to be exactly the same. The more we learn about the human genome, the more we uncover about the nongenetic influences that shape an individual. From its very first moments, the clone will be different from the DNA donor; the donor's nucleus will be put into an egg that is different from the egg in which the donor was conceived, which will affect the expression of the genome in unanticipated ways. Then the clone will be gestated in a different uterus and exposed to different maternal hormones and toxins, even to different sounds from the outside. And that's only the beginning; after birth, everything about the clone's environment will be different from the environment in which the donor was raised.

But even in the twenty-first century, people seem to believe that a clone would be fated to repeat the DNA donor's life — and would be tormented by the prospect. In her 2002 novel *The Secret,* for instance, author Eva Hoffman tells the story of a young woman and her beautiful, imperious, overprotective single mother. After asking for years who her father is, the young woman, Iris, finally discovers the truth: that she is the beautiful woman's clone. Her fury at her "mother" is immense. "She had taken it all away from me," Iris thinks: "my time, my place, my family and their need for me, my childhood, my slow latency, my prankish early teens. She'd been there first."

People imagine the particular anguish of the clone as something quite different from the way anyone else on earth, even an identical twin, would experience destiny and individuality. People once thought that the legacy of being a test tube baby would be just as unbearable. In vitro fertilization, many believed, would leave the child feeling the way Iris feels: like "a replica, an artificial mechanism, a manufactured thing." That didn't turn out to be the case, but the force of that conviction when IVF was new is well worth remembering today.

Should concern about creating such feelings in adult clones be sufficient to keep the science of cloning from getting under way?

Probably not. If we had banned IVF out of concern for how test tube babies would feel, we never would have had the blessing of half a million babies, children, and teenagers who basically feel just fine. And we would have limited what we could learn about fertilization, embryonic development, and early pregnancy, knowledge that will no doubt have some practical use in the end, but that is already precious just because it's knowledge.

More realistic fears about cloning focus on its medical rather than its psychological risks. Dolly the sheep died young, and according to some reports she was aging faster than sheep usually do. Other cloned mammals have developed a range of worrisome physical problems: umbilical cord defects, immunological deficiencies, placental abnormalities, and a birth defect known as large-fetus syndrome. It makes scientific sense that the reprogramming of an adult nucleus would come at a cost. Scientists already know that most cells have a finite life span, a fixed number of cell divisions carried in their internal clocks, and the adult cells responsible for Dolly and her ilk have already used up part of their quota before the cloning has even begun. It is because of health concerns like these, and because of a vague, generalized repugnance many people feel toward the idea of cloning, that anticloning laws and regulations have become the focus of so much debate.

For the past thirty years lawmakers and government officials have relied on expert panels to help them negotiate the most troubled waters of biomedical research. This has been called "commission ethics" — a term used as a subtle rebuke. Commission ethics is a crutch, some say, political cover to allow legislators or bureaucrats to take steps that might prove unpopular. Despite decades' worth of evidence to the contrary, politicians still hope that commissions will provide an easy solution, a quick way out of an ethical quagmire when an issue is emotionally or politically charged. But commission reports are almost always too complex for most regulators to read and fully understand. There are two reasons for this complexity, both related to the composition of the panels. The people named to bioethics commissions tend to be physicians, lawyers, and ethicists who have staked their professional reputations on exploring the what-if's of research; they love making things as complicated as they can. In addition, the mixture of viewpoints on these commissions is deliberately wide; most of the panels formed to review IVF research, for instance, have included both abortion opponents and IVF defenders, both theologians and humanists, both patient advocates and physicians. The only way to get all these contradictory opinions represented in a consensus document is to pepper it with caveats.

But politicians don't handle nuance well. They are usually stymied by commission reports, which are filled with qualifications: allow this but not that, permit this kind of research but only when it is conducted under that condition. Faced with the difficulty of turning a commission report into legislation, politicians often turn away from the report altogether. They end up doing nothing — or doing something blunt and simple, such as imposing an outright ban.

So far, few commission recommendations have been codified into law; one of the main weaknesses of commission ethics is the lack of regulatory teeth. The National Commission for the Protection of Human Subjects of Biomedical and Behavioral Research concluded in 1975 that fetal research was ethical, with restrictions; the Ethics Advisory Board said the same thing four years later about IVF research. Indeed, every ethics commission asked to consider this issue over the past thirty years has said that research on embryos, fetuses, and IVF is worthy of federal support — with limitations as to the age of the embryo or fetus, the length of time it is kept alive artificially, and the kind of informed consent obtained for its use. But a resistant executive branch, spurred by a conservative Congress and poked by the antiabortion lobby, has kept ignoring these recommendations and searching for a panel that would recommend something more to its liking. Each new commission is asked to reinvent the same wheel that has already been reinvented by the commission that went before.

In 1988 HHS directed NIH to put together an advisory committee to debate the ethics of fetal research — again. This time the matter came to public attention because of a much-touted new effort to treat degenerative brain diseases, such as Parkinson's, by transplanting tissue from aborted fetuses directly into patients' brains. The hope, which could not be substantiated without further research, was that the fetal tissue would take over for the damaged neurons and that the patient's body would not reject it because fetal tissue carried so little in the way of immunological identification. While the NIH committee, the Human Fetal Tissue Transplantation Research Panel, deliberated, HHS imposed a moratorium on such experiments. And when the panel issued its report, saying it was ethical — with certain safeguards — to conduct transplantation research with tissue from aborted fetuses, HHS ignored it.

So the 1988 moratorium on fetal research remained in effect for another five years, until President Bill Clinton signed into law the NIH Revitalization Act of 1993, which opened the door to federal funding for both fetal research and IVF research. The law specifically permitted NIH to fund studies of fetal cell transplantation, even though the therapy was already turning out to be less promising than scientists had hoped, and removed the requirement that IVF experiments be approved by a national ethics advisory board — a first step in allowing such research to receive federal funding at last, twenty years after Pierre Soupart first sent his IVF proposal over to the NIH.

Clinton had staked out a position on IVF and fetal research, but there was still embryo research to consider. For that the NIH Human Embryo Research Panel (HERP) was formed. In 1994 this group issued a report that echoed those of its predecessors: it was ethical to create embryos for research and to experiment on them up to the age of fourteen days. And like the earlier report, this one was ignored.

Next was the National Bioethics Advisory Commission (NBAC), appointed by Clinton in 1995. Among its sixteen members were physicians, lawyers, one patient advocate, one nurse, one psychologist, and one drug company representative. An indication of how entrenched bioethics had become by this point was that six of the members had job descriptions that included the word "ethics" — compared to just two when the first bioethics commission was formed more than twenty years earlier. NBAC had a broader mandate than did previous commissions: it was supposed to review all human subject research supported by every federal agency, not just NIH — and, at the same time, to address such knotty issues as gene patenting and genetic discrimination.

Almost as soon as Dolly the sheep was born in 1997, Clinton asked NBAC to write a report about human cloning, which it did that same year. In their deliberations, not only about cloning but about a variety of complex issues concerning the protection of human subjects, the commissioners kept returning to three principles that had guided bioethics ever since the first commission wrote its most influential paper, the Belmont Report, in 1975: the principles of autonomy, justice, and beneficence. As the NBAC chairman saw it, cloning could be judged according to these same principles — and it did not necessarily violate any of them. "There are no new ethical issues raised by reproductive cloning that are not raised by the use of genetic engineering and other reproductive technologies," the chairman, Princeton president Harold Shapiro, said after NBAC was disbanded and he was no longer speaking in an official capacity. "Ethically, it is a subset of these broader problems." The excessive concern about reproductive cloning, Shapiro said, amounted to "unjustified hysteria . . . much ado about nothing."

In 2001 President George W. Bush created yet another bioethics commission, this one to help him evaluate an issue that was causing him much public grief: whether to allow federal funding for embryonic stem cell research. Bush appointed Leon Kass, the IVF opponent who had already gone on record against cloning, to chair the President's Council on Bioethics. As its first order of business, the council decided to focus on the ethical issues raised by something tangentially related to embryonic stem cell research: the juiciest topic of the day, human cloning.

Throughout 2002 the council heard testimony about the pros and cons of cloning — scientific arguments, religious arguments, legal arguments, political arguments. Kass tried to broaden the debate by hiring outside consultants to write reports on such matters as whether research moratoriums ever worked. He also piqued everyone's interest — including the interest of the media, which was watching him closely — by assigning a short story to be read and discussed at the first meeting. The story was "The Birthmark" by Nathaniel Hawthorne, about a cocky young physician who is bothered by the single flaw in his otherwise perfectly beautiful wife, a tiny birthmark on her cheek in the shape of a hand. His attempt to remove it results in her death — making "The Birthmark" a story about the hazards of imposing man's yearning for perfection on an imperfect world. For subsequent meetings Kass occasionally included other provocative reading assignments, such as the essay "Whither Thou Goest" from *The Doctor Stories* by surgeon Richard Selzer. He also posted literature on the council's Web site: the poem "After Making Love We Hear Footsteps" by Galway Kinnell, for instance, and highlights from the script of the 1997 science fiction movie *Gattaca.*

Some members of the President's Council on Bioethics resisted treating the two forms of cloning, reproductive and therapeutic, as morally distinct. They forced the council to use terminology — "cloning-to-produce-children" and "cloning-for-biomedical-research" — that emphasized the similarities of the procedures rather than their differences. Many of the council's most conservative members, Kass included, were opposed to human cloning in any form. But they came to a compromise of sorts. In July 2002 the council report recommended an outright ban on cloning-to-produce-children, and a temporary moratorium, to be revisited in four years, on cloning-for-biomedical-research. In addition the group called for a federal review of "current and projected practices of human embryo research, pre-implantation genetic diagnosis, genetic modification of human embryos and gametes, and related matters," in the hopes that "ethically sound policies" would underlie decisions in the entire field of reproductive technology and genetic intervention.

"A world that practiced human cloning, we sense, could be a different world, perhaps radically different, from the one we know," the council concluded, using language as flowery as Kass's own in his anti-IVF articles from the 1970s. "It is crucial that we try to understand, before it happens, whether, how, and why this may be so." Creating cloned embryos, they wrote, "requires crossing a major moral boundary, with grave risks and likely harms, and once we cross it there will be no turning back."

Occasionally, the frightening predictions about where science might lead turn out to be every bit as terrifying as they had seemed in prospect. The atomic bomb is an obvious example. More often, the predictions turn out to be right in some, but not all, of their particulars. Like a portrait that captures someone's features but misses her spark of personality, most predictions don't tell the full story as it unfolds in real life. That was the case with test tube babies, who didn't have horns or tails or a third eye, who were just like every other baby conceived in more conventional ways — but whose existence did, as critics had feared, make possible other reproductive technologies that presented problems of their own.

Sometimes, rarely, predictions that come true end up with a hidden good side along with the bad, a benefit accompanying the risk. When this happens — as happened, for instance, with the genetically engineered mammal known as the transgenic mouse — it gives one pause about cutting off science too early based on fallible predictions about where it's all likely to lead.

When Jon Gordon and Frank Ruddle of Yale first inserted a bit of foreign DNA into a mouse embryo in 1980, and then grew the genetically engineered mouse, to show that the foreign DNA had been taken up by all of the mouse's cells, they were traveling down a path that critics had already warned could be socially disastrous. Opponents of gene swapping at the 1975 Asilomar conference had said it might lead to exactly the kind of research that Gordon and Ruddle were doing five years later. Ironically, however, by the time the Yale investigators proved the critics right, nobody seemed to notice. Had their work received wider publicity, there might have been newspaper columns in which career Cassandras warned that cross-species experimentation would lead to the creation of bizarre man-beasts reminiscent of H. G. Wells's *Island of Dr. Moreau.* How lucky it was, then, that Gordon and Ruddle published their work in a relatively obscure journal. They were allowed to continue their research, and no man-beasts were born. Eventually, the DNA that scientists inserted into their mice was indeed human DNA, but the result has been a benefit, not a bane. The transgenic mouse — a mouse with a touch of the human genome inside every one of its cells — has turned out to be one of the most valued research tools in biology today. It offers a revolutionary way to study biomedicine, to decipher the human immune system, to understand the workings of the human genome. The transgenic mouse has become, as one enthusiast puts it, "the *E. coli* of the twenty-first century."

Cloning might turn out to be like the transgenic mouse — a technology that skirts a dark and dismal future, leading instead to unexpected enlightenment. Making baby clones is not the only way to use the technique. Cloning can also perform functions that are not really frightening at all — for instance, making a customized egg for a woman who can't make eggs on her own. The scientist starts with a donated ovum, removes the nucleus, and inserts the nucleus from one of the woman's body cells. This is nuclear transfer, the first step in cloning. He then allows the egg to go through meiosis, after which it contains just one-half of the woman's genome, just as in an egg she had made herself. This new egg, with twenty-three chromosomes from the woman, can be fertilized with sperm from her husband, and the zygote will be just like any other IVF zygote, conceived in the laboratory by mixing genes from both a woman and a man. But the first step in making that zygote was the same step that would have been taken if the goal had been to make a clone.

Cloning now, like IVF then, stirs up basic feelings about the urge to reproduce, about identity, about what we expect from the next generation. Most of us are born with a drive to perpetuate ourselves; it's an animal hunger, hot and insistent, bubbling beneath the cool of rationality. This is why infertility can be so devastating, why well-meaning suggestions like "You can always adopt" often miss the mark. For many people, adoption won't quench the desire; the bond they're yearning for is a matter of genes. If there is a little piece of your own genome in your child, mixed ideally with your spouse's genome — but, if need be, even the genome of a stranger or of your wife's sister will do — you have managed to keep a part of you alive, in a way, forever. All the dreams and expectations, all the derailments and missteps of parenthood, carry on something of the person you are, because of that small biological bit. This explains, at least in part, why so many tens of thousands of couples have been willing to put themselves through so many tens of thousands of painful and expensive interventions to try to have babies of their own.

With cloning, the genome you're contributing is purely, completely yours, without the mixing that brings along so much uncertainty. But the purity of the genetic link is often beside the point. People who want to clone themselves don't necessarily want to create an identical copy; all they want is to be able to keep some of their own genome alive, and cloning happens to be the only way they can do it. The woman in premature menopause, who might be able to bear a child but who can't provide the eggs to do so, might see a chance at self-perpetuation in a clone using somebody else's egg and the nucleus of one of her own body cells. The man without a partner, who wants a son and heir and is willing to have a stranger bear one for him, might see cloning as a way to pass on his genome without having to add the surrogate mother's unfamiliar genes into the brew.

Reproductive technology, whatever else it can or cannot do, however else it threatens or ennobles us, can do one thing for sure. It allows a subset of individuals to transcend a mistake of nature — one that otherwise would have left them bereft — so that they, like the rest of us, can forge a link to the next generation.

23

Mixed Blessings

> Who controls the past controls the future; who controls the present controls the past.
>
> — George Orwell, *Nineteen Eighty-Four* (1949)

Doris del-zio never had another child. After she won the case against Raymond Vande Wiele in 1978, she became an active spokeswoman for in vitro fertilization. She signed on to a three-city television tour sponsored by the Westinghouse Group, promoting assisted reproduction options for women. She sold her story for $5,000 to *Good Housekeeping*, which ran an article in March 1979 under her byline called "I Was Cheated of My Test Tube Baby." She gave a few talks to small groups in Fort Lauderdale and elsewhere in Florida. It was strange to see this traditional, conventionally religious, naturally reticent woman publicly aligning herself with a technology that many other traditional, conventionally religious people considered a dangerous intrusion into sacred territory.

Then, just as quickly as she entered the public arena, Doris ducked out. She stopped accepting speaking engagements, having decided that she no longer wanted to replay the pain of September 13, 1973, no matter how important the cause. She stopped talking about yearning for something she didn't have and focused instead on what she did have: her daughter, her daughter's three children, her new career as a bank officer. John Del-Zio, meanwhile, toyed with the idea of writing a book about the couple's misadventures with IVF. After retiring from dentistry, he enrolled in adult education classes in creative writing to prepare himself for the project. He stuck with the classes, but nothing ever came of the book.

John and Lesley Brown, in contrast, did write a book. In 1979, with the help of a professional writer, they published *Our Miracle Called Louise.* It was written in a chatty style, but it was honest, too; it pulled no punches about John's philandering and Lesley's less-than-pristine behavior as a teenager. The Browns grew accustomed to making declarations about their little miracle. On every landmark birthday of Louise's — when she turned one and two and three, when she turned ten, twenty, twenty-one — quotes from John and Lesley and, as she got older, from Louise herself would appear in newspapers all across Great Britain. On her tenth birthday in 1988, Louise announced that she wasn't a test tube baby anymore; "I am a test tube big girl now." On her twentieth birthday in 1998, a party was given in her honor in the House of Commons, attended by the British minister of health and by a member of Parliament who had had a baby by IVF herself. Twenty-year-old Louise, blond and curly-haired, chubby and soft-spoken, worked as a nursery school aide in Bristol and liked to spend evenings playing darts.

When Louise turned twenty-one, her younger sister, Natalie, then seventeen, made some medical history of her own. With the attention of the world's press already focused on the Browns because of Louise's birthday, Natalie announced that she had recently become the first test tube baby to give birth to a baby herself. Natalie's daughter, Casey, born May 13, 1999, had been conceived the old-fashioned way — further proof that test tube babies were just like anybody else.

Landrum Shettles left New York soon after resigning from his position at Columbia. He moved to Randolph, Vermont, in 1975 and became director of obstetrics and gynecology at Gifford Memorial Hospital. There he continued with his solitary research, once again working in a makeshift laboratory far into the night. At the time, he was studying what is known as a giant spermatozoon, a larger-than-normal sperm that has failed to undergo meiosis, meaning that it still has its regular complement of forty-six chromosomes. The giant spermatozoon occurs spontaneously about 1 or 2 percent of the time.

If a giant spermatozoon penetrates an egg, as occasionally happens, it's as if two sperm had done the fertilizing, resulting in a triploid zygote that is destined to die. But Shettles used the anomalous giant for his own purposes. He removed the nucleus from a human egg and inserted instead the nucleus of a giant spermatozoon. Then he grew the egg in the laboratory for a few days, letting it split into two cells, then four, then sixteen. What he had, in effect, was a human clone — an embryo whose entire genome derived from a single parent, in this case the man who provided the giant spermatozoon. When the administrators of Gifford Memorial found out what Shettles was doing, they were horrified. They ordered him to stop immediately and said that if he didn't, they would fire him.

Shettles did stop, but he was frustrated and angry. Soon he was approached by the directors of a quirky research foundation who wanted to set up a cloning clinic in the desert near Las Vegas. Cloning was not illegal in Nevada — nor would it ever be, the clinic backers believed, since this was a state with little taste for regulation, a state where both gambling and prostitution were permitted. Shettles accepted the offer to head the cloning clinic and moved to Las Vegas in 1981. The clinic never took off, but he stayed in Las Vegas until 2001, when he was no longer able to care for himself. He ended up in a nursing home in Florida, visited every day by his oldest daughter, Lana, who said she never really knew her father until he was an old man. He died in February 2003 at the age of ninety-three.

Raymond Vande Wiele, who died in 1983, did not live long enough to see in vitro fertilization become an ordinary thing. Nor did Patrick Steptoe, who died in 1988 at the age of seventy-five. William Sweeney, Shettles's collaborator, was also seventy-five when he died in 1997, struck by a heart attack in his office while he was seeing a patient.

Howard Jones lived into the twenty-first century and watched all the new developments in assisted reproduction with fascination and, occasionally, alarm. Even in his nineties he drove to work every day, flew to international conferences, and continued to write articles and books. Georgeanna accompanied him on his daily trips to the fertility clinic that is now named the Jones Institute, but by 2001 she had developed severe Alzheimer's disease. She sat quietly at her desk in the office adjoining Howard's, arranging piles of papers and waiting for her husband to take her home.

Howard kept a special spot in his heart for Elizabeth Carr, his first IVF success, the girl who almost shared his birthday. Every December 28 he would call her to wish her happy birthday and see how she was doing. And Elizabeth, who said she always thought of the Joneses as another set of grandparents, was healthy and happy, an adored only child with shiny dark hair and a dazzling smile. The *Today Show* and the *New York Times* and *Glamour* magazine phoned her periodically to discuss cloning or other controversies about which America's first test tube baby was expected to have an opinion. After being interviewed by so many reporters who hadn't done their homework, Elizabeth said, she decided to go into journalism, preferably television news, in hopes of doing it better.

Robert Edwards, one of the youngest of the IVF Mafia dons in the early days, entered the twenty-first century filled with new ideas. In his late seventies he was still giving lectures around the world, maintaining the frenetic tempo of travel he had danced to his whole life. In 2001 he inaugurated an electronic scientific journal, RBM Online, that reported on trends in reproductive technology. That year, when questions about the ethics of embryonic stem cell research made headlines in both the United States and Britain, Edwards posted an article saying it was his research that had made stem cell cultures possible in the first place. The claim may have been a stretch, but it did raise one important point: almost all of the hot spots of genetics research in the twenty-first century had their origin in the IVF research conducted thirty years before.

James Watson, the Nobel laureate who had blasted Edwards for his IVF work in 1971 and who had raised the specter of "clonal man" as something to be feared, seemed in his old age to grow more sanguine about altering the genetic destiny of humans. In the fall of 2002 Watson, then seventy-four, told a reporter that it would be not only acceptable but admirable to use our knowledge of the human genome to make better babies. He said it would be a good thing to manipulate genes to enhance a child's personality, intelligence, or physical qualities. In his inimitable way, Watson reduced the matter to its starkest terms: "Who wants an ugly baby?"

Science is a Janus-faced creature: it can elevate us, or it can bring us to grief. Through science we learn not only how the world works — how the universe began, how genes are expressed, how species become extinct — but also how we can use such insight to better ourselves. Science can also endanger us, however; without theoretical physics, there could have been no atomic bomb.

Both sides of scientific inquiry, the light and the dark, revealed themselves in IVF research. When scientists learned to create and culture human zygotes in a petri dish, they gained a new understanding of embryonic development and differentiation, which led in turn to improved prenatal care and the prevention of some birth defects. The technology also provided a benefit that was immediate and clear: babies for infertile couples who desperately wanted them. But IVF also provided the laboratory techniques that made possible the creation of startling new forms of life: human-animal chimeras, made-to-order babies, human clones. In order to reap the many benefits of IVF, we have had to learn — we are learning still — how to limit the applications that might turn out to be the most dreadful.

The story of the acceptance of in vitro fertilization shows that the slope can indeed be slippery. When scientists step into unknown territory, they risk sliding down a perilous path. But there is another side to that slope: many new discoveries will carry us upward rather than down, off to somewhere better than where we began.

We can't always know in advance which way a particular discovery will take us. So we should probably hold off on wrestling our demons until we find out exactly what they are, rather than taking steps to avoid them based on the shadow they cast in our imaginations. Often the demons are more ominous in prospect than they turn out to be in fact. Many of the predictions about where IVF would lead, for example, never really came to pass. One scenario was that it would bring us wombs-for-hire, an oppressed underclass of women paid to bear the children of the infertile rich. But surrogate motherhood is expensive, unpleasant, and emotionally complex, and it never became all that popular. Another potential monster was human cloning, but even cloning may turn out to be less prevalent and less scary than we imagined. Market forces might make reproductive cloning impractical, and scientific advancements may make it unnecessary. In the future, the term "cloning" might refer only to therapeutic cloning, and it might someday become a real therapy — a laboratory technique for creating cells to regenerate damaged organs, perhaps, or for manufacturing artificial eggs for women who can't make their own.

One thing above all accounts for our deepest fears about reproductive genetics: we don't know how the story will end. We can't predict the shenanigans still ahead; we can't predict what will become of the clones or the chimeras or the genetically enhanced. Nor can we predict the outcome for the rest of us if such tinkering is allowed. But in the 1970s we didn't know how the IVF story would end, either. In vitro fertilization was frightening because at the time anything seemed possible, the worst outcome every bit as likely as the best. Only today, against a backdrop of half a million beloved and generally healthy test tube babies, can we see that the IVF story basically had a happy ending — hundreds of millions of human gametes have been handled in the lab, and life goes on. Will the other stories have happy endings, too?

In the final analysis, we must act without knowing what will happen next; the urge to act, to do whatever it takes to find things out, is embedded in our human core. We have the curiosity of Pandora, opening whatever boxes we encounter, no matter what terrible secrets we might reveal along the way. But we have arrogance, too, the arrogance of Prometheus, Icarus, or Oedipus, mythological figures who met their doom while trying to outsmart their tattered destinies. It's treacherous, this striving to transcend our fate, and it could lead to repercussions worthy of the best Greek tragedy. We risk it anyway. Because we believe that the good outcome is just as likely: that we will adapt to new discoveries the way we have so often adapted, incorporating them into the shifting terrain of genomes, genes, and generation.

Notes

Prologue: Monster in a Test Tube

1 **"It was a familiar routine":** Doris Del-Zio's medical history is from a deposition taken in preparation for the trial of John E. Del-Zio and Doris Del-Zio, Plaintiffs, v. The Presbyterian Hospital in the City of New York, Raymond L. Vande Wiele and the Trustees of Columbia University in the City of New York, Defendants (case no. 74 CIV 3588), hereinafter referred to as "the Del-Zio trial." The case was filed in August 1974 in the United States District Court for the Southern District of New York and was finally brought to trial on July 17, 1978. The trial documentation — which includes 4,500 pages of trial transcripts, depositions, plaintiff's and defendant's exhibits, charges to the jury, and appeals — is stored in seven large cardboard boxes at the National Archive facility in a suburb of St. Louis. These boxes were transferred at my request to the National Archive's New York City branch in downtown Manhattan, where I could inspect and photocopy them.

2 **"They can put a man on the moon":** deposition of Doris Del-Zio, Del-Zio trial, July 1975, pp. 36–37.

width of an eyelash: Roger A. Pedersen, a specialist in embryonic development at the University of California at San Francisco, quoted in Earl Lane, "Experiment with Embryos; Panel Suggests New Rules," *Newsday*, Sept. 27, 1994, p. 27.

3 **"of our own free will":** plaintiff's exhibit no. 18 in the Del-Zio trial.

4 **eighth-floor operating room:** interview with Ionnis Zervoudakis, August 2002. Zervoudakis, who was a third-year resident in 1973, assisted Sweeney in the operation on Doris. Now a New York Hospital obste-

tričian/gynecologist, he told me that the eighth-floor operating room had a beautiful view of the East River and Roosevelt Island.

4 **Sweeney collected:** as described by Landrum Shettles in a summary written on Oct. 11, 1973, at his lawyer's request. It was offered in evidence at the Del-Zio trial five years later.

6 **"Marriage and the family":** quoted in Bruce J. Schulman, *The Seventies: The Great Shift in American Culture, Society, and Politics* (New York: Da Capo Press, 2002), p. 165. Atkinson's essays are collected in *Amazon Odyssey* (New York: Links Books, 1974).

9 **"Thy godlike crime":** lines 35–38 of "Prometheus," published in 1816. It can be found on-line at www.library.utoronto.ca/utel/rp/poems/byron8.html.

10 **"A blastocyst":** Leon Kass, "'Making Babies' Revisited," *Public Interest,* Winter 1979.

"Fecundation must" and "we would not like": quoted in Mark Liff, "Church Denounces Test Tube Birth," *New York Daily News,* July 28, 1978, p. 3.

11 **human embryos explicitly:** The research was conducted at the Jones Institute for Reproductive Medicine in Norfolk, Virginia, where the first American test tube baby had been born in December 1981. See Sheryl Gay Stolberg, "Scientists Create Scores of Embryos to Harvest Cells," *New York Times,* July 11, 2001, p. A1; Rick Weiss, "Scientists Use Embryos Made Only for Research," *Washington Post,* July 11, 2001, p. A1; and Stolberg, "For Clinic, Stem Cell Test Is Rebirth of Old Debate," *New York Times,* July 12, 2001, p. A19.

14 **"If any one age":** Quoted in Leon Kass's chapter in Michael Hamilton (ed.), *The New Genetics and the Future of Man* (Grand Rapids, Mich.: William B. Eerdmans, 1972), p. 50.

15 **"black stinking cloud":** This version of the Pandora's box story can be found on-line at www.geocities.com/una_sorella/stories/pandora.htm.

1. Room Temperature

19 **he asked the woman:** testimony of Landrum Shettles, trial transcript, p. 2316.

"a larger version": as remarked by John J. Bower, the attorney for Columbia University during the Del-Zio trial, in the judge's chambers, trial transcript, p. 9.

"we'll keep you informed": John Del-Zio's deposition, which was read in the courtroom, trial transcript, p. 2489.

20 **as if a novelist:** Norman Mailer, *The Prisoner of Sex* (New York: New American Library, 1971), p. 146. Also quoted in David M. Rorvik, "The Penthouse Interview: Dr. Landrum Shettles," *Penthouse*, May 1981, p. 133.

21 **"to try to find out":** Del-Zio trial transcript, pp. 1275–76. The list of cities he visited is among his personal papers, which are in the possession of his daughter, Lana Shettles Callahan of St. Petersburg, Florida, whom I visited in November 2002.

22 **"persevering and visionary":** letter of recommendation dated Nov. 15, 1973, one of a dozen or more such letters that Shettles apparently solicited and collected after resigning from his Columbia position in October 1973. They are in the possession of Lana Shettles Callahan.

"sheer enthusiasm": personal interview with Howard Jones, January 2002.

"we have invested": in Gebhard F. B. Schumacher, ed., "In Vitro Fertilization of Human Ova and Blastocyst Transfer: An Invitational Symposium," *Journal of Reproductive Medicine* 11, no. 5 (Nov. 1973): 197.

23 **He pulled out:** Shettles's summary of the procedure, Oct. 11, 1973.

frothy chocolate milkshake: interview with Allerand, October 2001. She was a little less colorful in her testimony at the Del-Zio trial, saying "it was a horrible-looking mess" that "looked like an abscess which had been aspirated," trial transcript, p. 2373.

24 **"Mrs. Egg":** interview with Jagiello, May 2001.

his greatest passion: Vande Wiele's likes and dislikes and his style of entertaining were described by his friend Michel Ferin in an interview in May 2001.

25 **the ugly truth:** On July 15, 1986, years after he left Columbia, Shettles wrote to the famed Nazi hunter Simon Wiesenthal in Vienna to check out rumors about the Vande Wieles. Wiesenthal wrote back on Aug. 5 that Vande Wiele's older brother, Dr. Jos Vande Wiele, was in the Flemish Legion and had attained the rank of Obersturmbannführer in the Nazi party on December 9, 1944. He said he had no information about Raymond Vande Wiele.

profoundly enraged: Michel Ferin remembered how furious Vande Wiele was with Shettles on Sept. 13 in my interview with him in May 2001.

25 **Vande Wiele called:** Shettles's summary written on Oct. 11, 1973. He said that Vande Wiele had phoned Mr. Alvin J. Binkert, the president of the hospital, and Dr. Donald Tapley, the acting dean of the medical school, and had requested an emergency meeting of the department executive committee for the following Monday, Sept. 17. Then Vande Wiele had called Shettles himself.

2. The Dance of Love

26 **It began when:** The story of William Pancoast's artificial insemination was first reported by one of his students, Addison Davis Hard, in an article written twenty-five years after the fact — and eleven years after Pancoast's death — for *Medical World.* The case is described in Robert T. Francoeur, *Utopian Motherhood: New Trends in Human Reproduction* (Garden City, N.Y.: Doubleday, 1970), pp. 1–4.

28 **"same race stock":** the quotes are from an article in *Journal of the American Medical Association* (hereinafter abbreviated as *JAMA*), which was summarized in the unsigned "Test-Tube Babies," *Literary Digest,* Nov. 21, 1936, p. 19.

Doornbos v. Doornbos: The case (23 U.S.L.W. 2308, Ill. Super. Ct. Dec. 13, 1954) is described in J. M. Shaman, "Legal Aspects of Artificial Insemination," *Journal of Family Law* 18 no. 331 (1979–80).

Feversham Report: see "Human Artificial Insemination: Feversham Committee's Report" (editorial), *British Medical Journal,* July 30, 1960, pp. 379–80; and "Artificial Insemination" (editorial), *Lancet,* July 30, 1960, pp. 247–48.

29 **Pincus mixed:** Howard Jones, phone conversation, March 2003. Jones did not think Pincus's experiment qualified as "laboratory fertilization" — it was more like the IVF variation known as GIFT (gamete intrafallopian transfer). See G. Pincus and E. V. Enzmann, "Can Mammalian Eggs Undergo Normal Development in Vitro?" *Proceedings of the National Academy of Sciences* 20 (1934): 121–22.

"if such an accomplishment": "Conception in a Watch Glass," *New England Journal of Medicine* (hereinafter abbreviated as *NEJM*) 217, no. 17 (Oct. 21, 1937): 678.

30 **anonymous author:** The theory that John Rock wrote the *NEJM* editorial was offered in Nicola Perone, "In Vitro Fertilization and Embryo Transfer: A Historical Perspective," *Journal of Reproductive Medicine* 39

(1994): 698. That seems a reasonable guess, considering the jaunty, jokey tone of Rock's letters to Shettles (in Shettles's private collection) and how well it matches the tone of the editorial.

the first successful fertilization: John Rock and Miriam F. Menkin, "In Vitro Fertilization and Cleavage of Human Ovarian Eggs," *Science* 100, no. 2588 (Aug. 4, 1944): 105–7.

scientists were skeptical: Perone, "In Vitro Fertilization," pp. 695–700.

31 **"a monstrosity":** AP, "Scientists Fertilize Human Egg in Test Tube," *Daily American,* Jan. 14, 1961. The research is also described, without citation, in Paul Ramsey, "Shall We 'Reproduce'?" *JAMA* 220, no. 10 (June 5, 1972): 1347; and in Francoeur, *Utopian Motherhood,* p. 57.

"monstrous": AP, "Vatican Calls Test Tube Baby 'Immoral,'" *San Francisco Examiner,* Jan. 14, 1961, p. 6.

But it was the way: Francoeur, *Utopian Motherhood,* p. 59. The newspaper *Paese Sera* was mentioned by name in "Vatican Calls Test Tube Baby 'Immoral.'"

32 **Moscow scientists claimed:** Francoeur describes Petrucci's work with the Russians in *Utopian Motherhood,* p. 58. I have never seen any other mention of the experiment, which Francoeur said was conducted in 1966 by Pyotr Anokhin and Ivan Nikolaivitch Maiscki.

"When they visited": Robert Edwards and Patrick Steptoe, *A Matter of Life: The Story of a Medical Breakthrough* (New York: William Morrow, 1980), p. 38.

33 **Only one physician:** *ibid.,* p. 40.

something went wrong: The same problem occurred in other species, including mice and cows in Edwards's own laboratory. Edwards describes the problems with lab-ripened eggs of all species in *A Matter of Life,* p. 87.

34 **60 million:** That number, arrived at in 1979 and, compared to 90 million estimated in 1929 — is from Lori Andrews, *The Clone Age: Adventures in the New World of Reproductive Technology* (New York: Henry Holt, 1999), p. 18. Clifford Grobstein, in *From Chance to Purpose: An Appraisal of External Human Fertilization* (Reading, Mass.: Addison-Wesley, 1981), says that a normal male ejaculate has 50 to 60 million sperm per milliliter and that in vitro only about 50,000 sperm are needed to ensure that one will penetrate the egg (p. 69).

human sperm that reach: Robin Marantz Henig, "In Vitro Fertilization: A Cautious Move Ahead," *BioScience* 28, no. 11 (Nov. 1978): 685–88.

possibly undermining: Concerns about fertilization by multiple sperm or abnormal sperm were expressed by John Biggers of Harvard, quoted in Dorothy Nelkin and Chris A. Raymond, "Tempest in a Test Tube," *The Sciences* 20 (Nov. 1980): p. 6 ff.

35 **"egg poacher":** David M. Rorvik, "Your Baby's Sex: Select, Don't Settle," *New York,* May 19, 1969, p. 37.

"the dance of love": quoted in David M. Rorvik, "The Test-Tube Baby Is Coming," *Look,* May 18, 1971, p. 83 ff.

in 1962: The actual date of this supposed embryo transplant is difficult to track down. Shettles later claimed to have described the embryo transfer in "Einbettung und Anseidlung des Menschlichen Embryo," *Medizinische Klinik* 58 (Oct. 1963): 23–24. He mentioned the German publication in a letter of May 20, 1971, to Raymond Vande Wiele, which was admitted as evidence during the Del-Zio trial. In it he says that the direct quotes from him in Rorvik, "Test-Tube Baby Is Coming," "appear to be taken from" the German journal article. Rorvik, however, mentioned two different dates for the embryo transfer in two different articles. In *Look* he said it occurred in 1961; in "The Embryo Sweepstakes," *New York Times Magazine,* Sept. 15, 1974, p. 17, perhaps mistaking the date of the German publication for the date of the operation itself, he said it occurred in 1963. To add to the confusion, Shettles himself sometimes said that it occurred in 1962, sometimes in 1964. Throughout the Del-Zio trial he claimed it was 1962, probably during the second half of the year (see, for instance, his testimony under cross-examination, transcript, pp. 1612 and 1616–17).

But in a letter to the editor of *Ob-Gyn News,* Sept. 15, 1976, p. 4, in which he expressed skepticism about Edwards and Steptoe's claim that they had accomplished the first documented IVF pregnancy the previous June, Shettles said that he had written up his 1964 experiment in one of the supplements to his book *Ovum Humanum: Growth, Maturation, Nourishment, Fertilization, and Early Development* (New York: Hafner, 1960), though he failed to give a date for the supplement. To further complicate things, when one of the defense lawyers combed through hospital records of Columbia Presbyterian to confirm Shettles's claim for the implantation operations and the hysterectomy, no record of them was found in the surgical logs for all of 1962. However, according to a long letter about the trial written by prosecuting attorney James Furey to Eugene T. DeCanio of the Hartford Insurance Company, two operations conducted there in December 1961, five days apart, "had some differences and some similarities with the description given by Dr. Shettles. Under the circumstances, we did not bring in the surgical logs." So which date of the four possible dates should one believe? I chose 1962, since that was the date offered by Shettles himself under oath.

36 **Shettles may have told:** James Furey recalled, in an interview in May 2001, that Shettles testified at the Del-Zio trial that he had never told this woman anything, had not told his resident what he was doing, had not taken any notes, and could no longer remember the patient's name. Trial transcript, pp. 1608–19.

she was thirty-six: Shettles, letter to the editor, *Ob Gyn News*, Sept. 15, 1976, p. 4.

using a pipette: "In Vitro Fertilization," *Journal of Reproductive Medicine* (1973); in his letter to the editor of *Ob Gyn News* (1976), he said that in addition to the segment of caudal catheter, he used a tuberculin syringe.

37 **he saw no reason why:** "In Vitro Fertilization," p. 200; if not for the hysterectomy, he concluded, the world's first IVF pregnancy might well have proceeded normally, since "no contraindication for continued development was discernible." Shettles did not inform his peers about this remarkable feat until eleven years later, two years after it had been reported in Rorvik's *Look* article. However, in his letter of May 20, 1971, to Vande Wiele, Shettles wrote that he had already reported the experiment in the German peer-reviewed journal *Medizinische Klinik.* After looking up the article, however, Vande Wiele wrote back on June 4 that in it, "as far as I can make out, and my German is still not very rusty, there is no mention of any manipulations."

same hallway refrigerator: interview with James Furey, May 2001.

"The publicity": A. McLaren and M. G. Kerr, "Egg Transplantation," *Science Journal,* June 1970, p. 56, cited in Leon Kass's chapter in Hamilton, *New Genetics and the Future of Man,* p. 20.

38 **"Life is uniquely":** Robert Edwards, *Life Before Birth: Reflections on the Embryo Debate* (New York: Basic Books, 1989), p. 2.

39 **stunning coincidence:** "There is an internal problem with Shettles's account," Howard Jones wrote to me in March 2003. "Implantation is a process that requires time. Therefore for the embryo to have 'implanted' is really not credible. Furthermore, implantation requires precise embryo-endometrial synchrony, a coincidence not likely on a

chance basis." I have tried to indicate the uncertainty of Shettles's claims by making it clear that no one but Shettles himself and his frequent popularizer, David Rorvik, ever made them.

39 **scientific progress:** See "Showcasing Technology at the 1964–1965 New York World's Fair," naid.sppsr.ucla.edu/ny64fair/map-docs/technology.htm.

40 **"mouse geneticist":** Howard Jones told this joke at the Serono Symposium held in his and Georgeanna Jones's honor, "Optimizing ART," Apr. 4–7, 2002, in Williamsburg, Virginia.

41 **Each time Jones:** Jones's daughter, a student at Mt. Holyoke College who was working in the operating room that summer, was enlisted to remind physicians that "ovarian tissue from young patients was much appreciated" (Howard Jones, "In the Beginning There Was Bob," *Human Reproduction* 6, no. 1 [1991]: 5). She recalled this during an interview in April 2002.

Capacitation drove Edwards: Edwards and Steptoe, *Matter of Life*, p. 54.

42 **He often accompanied:** Howard Jones recalled their sailing days in an e-mail message of Oct. 1, 2002. "The sailboat did have a name," he wrote. "It was called *Aphrodite* and it had a dinghy called *Hermaphrodite.* We sailed in the Chesapeake Bay for the most part, although we sometimes went into the Delaware Bay and on up to New York."

"a kind of fairy-tale marriage": Details about the Joneses' personal and professional lives, as well as about their marriage, come mostly from the toasts and testimonials presented during a three-day Serono Symposium sponsored by the Bertarelli Foundation held in their honor in April 2002 in Williamsburg, Virginia. Among the attendees whom I interviewed privately during the conference were those who had known the Joneses for decades: Robert Edwards, Zev Rosenwaks, Lucinda Veecks, and their daughter Georgeanna Klingensmith.

44 **busy gynecological practice:** Edwards and Steptoe, *Matter of Life*, p. 67.

One technique he worked hard: *ibid.*, p. 68.

46 **"if the knowledge":** unpublished transcript in the archives of the Joseph P. Kennedy, Jr., Foundation, Washington, D.C. Shriver's speech was written up in Martin Tolchin, "Mrs. Shriver Gets a Lasker Award: Questions Letting Scientists Control Human Behavior," *New York Times*, Nov. 18, 1966.

ethical standards: Barbara J. Culliton, "National Research Act: Restores Training, Bans Fetal Research," *Science* 184 (Aug. 2, 1974): 426–27.

47 **"Someday, after mastering":** transcript of Shriver speech.

He had implanted: "The First Heart Transplant," from the Web site of the Groote Schuur Hospital, where Barnard worked, www.gsh.co.za/ab/heart.html. A more interesting, and critical, article about the event appears on the Web site "Icons of the Twentieth Century," www.abilene2000.com/icons/0928.html. Barnard's and Washkansky's comments are quoted in Peter Hawthorne, *The Transplanted Heart* (Johannesburg: Hugh Keartland, 1968), pp. 58, 132. Barnard later said he had been stunned by the international clamor. He had thought the procedure so noncontroversial that he had not even told the Groote Schuur administrators what he was doing. His recollection of the surgery was quoted in an online obituary posted by the BBC, at news.bbc.co.uk/1/hi/health/1470356.stm, when Barnard died in 2001 at the age of seventy-eight.

48 **"tender heart":** quoted in Hawthorne, *Transplanted Heart*, p. 149.

49 **"Nonsense!":** Edwards and Steptoe, *Matter of Life*, p. 77.

50 **In 1969 a Harris poll:** The poll, published in *Life*, June 13, 1969, is described in Francoeur, *Utopian Motherhood*, pp. 114–19.

"It is the implications": quoted in Edwards, *Life Before Birth*, p. 8.

51 **"I am one":** Edwards and Steptoe, *Matter of Life*, p. 89.

"You certainly started": John Rock to Landrum Shettles, Jan. 5, 1970. The letter is in the possession of Lana Shettles Callahan.

52 **One photograph:** Shettles, *Ovum Humanum*, p. 73.

more than one thousand: *ibid.*, p. 78, "Materials and Methods."

Even the archdiocese: The letter requesting permission, dated Sept. 18, 1963, and sent to Shettles at Presbyterian Hospital, is in the possession of Lana Shettles Callahan.

"The question of human": Pope Paul VI, *Humanae vitae, Acta Apostalicae Sedis* 60 (July 25, 1968): 388–89.

53 **"is a chance composite":** Roberts Rugh and Landrum B. Shettles with Richard N. Einhorn, *From Conception to Birth: The Drama of Life's Beginnings* (New York: Harper & Row, 1971), p. viii.

54 **With the Shettles technique:** Landrum Shettles and David M. Rorvik, *How to Choose the Sex of Your Baby* (reprint of *Your Baby's Sex:* New York, Doubleday, 1997) p. 4.

"flat on her back": Rorvik, "Penthouse Interview."

"I have none": David M. Rorvik, "Your Baby's Sex: Select, Don't Settle," *New York*, May 19, 1969, p. 37.

55 **embryo in aspic:** Information about how the embryo was prepared for photographing is from Gerald Leach, *The Biocrats: Ethics and the New Medicine* (Baltimore: Penguin Books, 1970). Leach used the photo on the cover of his book.

3. Laughingstock

57 **He did not have the heart:** Del-Zio trial transcript, p. 3434.

Tropical fish tanks: This description of Shettles's hospital office is from my interview with his daughter Lana Shettles Callahan, Nov. 5, 2002.

58 **"think again":** according to Georgianna Jagiello in an interview on June 4, 2001.

The primary guide: World Medical Association, *Declaration of Helsinki*, adopted by the Eighteenth World Medical Assembly, Helsinki, June 1964 (later amended by the Twenty-ninth World Medical Assembly, Tokyo, Oct. 1975; by the Thirty-fifth World Medical Assembly, Venice, Oct. 1983; and by the Forty-first World Medical Assembly, Hong Kong, Sept. 1989).

59 **"any medical or surgical":** quoted by Vande Wiele in a letter to Landrum Shettles dated Sept. 14, 1973, and admitted into evidence in the Del-Zio trial. Vande Wiele underlined the phrase "any withdrawal or removal of body tissues or fluids."

Another mechanism: John D. Roberts, a professor of internal medicine at Virginia Commonwealth University, described the content and function of a "letter of assurance" for the course Scientific Integrity, in a syllabus posted online at www.courses.vcu.edu/rcr/course_materials/hum-lec.htm.

60 **"It is annoying enough":** letter to acting chairman Charles M. Steer, June 12, 1969. The letter was a plaintiff's exhibit at the Del-Zio trial and read in court, transcript, p. 1362.

"A number of claims": the statement to the press, dated Mar. 21, 1972, was a plaintiff's exhibit at the Del-Zio trial.

61 **"I am sorry":** In Edwards and Steptoe, *Matter of Life*, p. 106.

Between 1970 and 1980: NCI and NIH budget information can be found at nih.gov/about/almanac/appropriations/index.htm.

62 **Gabriele Fallopio:** (sometimes spelled Gabriel Fallopi). See Simon Fishel, "IVF — Historical Perspective," in Fishel, ed., *In Vitro Fertiliza-*

tion: Past, Present, Future (Washington, D.C.: IRL Press, 1988). According to Howard Jones in a note to me in March 2003, whereas scientists in the 1970s thought the embryo did not reach the uterus until the blastocyst stage, "we now understand that it is the morula which usually reaches the uterus."

62 **"its trip down the oviduct":** Grobstein, *From Chance to Purpose,* pp. 38–40.

63 **To fertilize rabbit eggs:** M. C. Chang's research on sperm capacitation in rabbits, as well as mice, hamsters, and rats, is summarized in Jean L. Marx, "Embryology: Out of the Womb — Into the Test Tube," *Science* 182 (Nov. 23, 1973): 811–14.

Hamster embryos: see the chapter "Artificial Fertilization" in Alun Jones and Walter Bodmer, *Our Future Inheritance: Choice or Chance?* (London: Oxford University Press, 1974), pp. 30–44.

"kill you professionally": Vande Wiele's letter to Shettles, dated May 12, 1971, was an exhibit in the Del-Zio trial.

4. Out of Control

64 **It featured the premiere:** information about Bernstein's *Mass* and opening night can be found at the official Leonard Bernstein Web Site, www.leonardbernstein.com/studio/element.asp?FeatID=12& AssetID=24.

65 **"coded messages":** *ibid.*

Who should survive?: Harold M. Schmeck, Jr., "Parley Discusses Life-Death Ethics," *New York Times,* Oct. 17, 1971, p. 33.

66 **JFK's older brother:** Interesting tidbits of information about the Kennedy clan can be found at www.thejfkstamps.com/clan.asp.

first university-based institute: Stuart Auerbach, "GU to Study Medicine's Life and Death Decisions: Institute for 'Bioethics,'" *Washington Post,* Oct. 2, 1971.

"to fill the need": The Hastings Center's declaration of purpose appears on the inside front cover of each issue of the *Hastings Report.* Cofounder Daniel Callahan was invited to the October 1971 conference of the Kennedy Foundation as one of three "essayists" for the panel "Who Should Be Born: Is Procreation a Right?" Also on the panel were James Crow, a geneticist at the University of Wisconsin, and John Noonan, a professor of law at Berkeley. The session was moderated by John Chancellor of *NBC News,* and the experts responding to the panel's three essays were Joshua Lederberg, professor of genetics at Stanford; Jerome LeJeune, professor of genetics at the University of Paris; Claudine Escoffier-Lambiotte, medical editor of *Le Monde;* and American author William Styron.

67 **the year's surprise bestseller:** As of 1993 Richard Bach's *Jonathan Livingston Seagull* had sold an estimated 30 million copies in thirty-six languages, according to S. J. Diamond, "Sequel/'Phenomenon' Authors; Singular Sensations," *Los Angeles Times,* Feb. 1, 1993, p. E1.

"Without answers": quoted in Colman McCarthy, "Can Science and Ethics Meet?" *Washington Post,* Oct. 18, 1971.

68 **a Harris poll revealed:** as reported in *Life,* June 13, 1969, p. 52. The poll is also mentioned in Edward Grossman, "The Obsolescent Mother: A Scenario," *Atlantic,* May 1971, p. 47, and in Francoeur, *Utopian Motherhood,* p. 57. The estimate of the birth rate from artificial insemination is from *Utopian Motherhood,* p. 19.

he kept no lab notebooks: Del-Zio trial transcript, pp. 1608–15.

69 **"The crowd was modish":** as described in the transcript of "Men and Molecules," a radio show sponsored by the American Chemical Society and narrated by Stuart Finlay, script no. 569 in the archives of the Joseph P. Kennedy Jr. Foundation.

The second was Howard Jones: Jones offered his theory about why he was invited in a phone conversation in March 2003.

70 **an interest sparked:** Kass has said that he was motivated to switch from biology to philosophy by a column in the *Washington Post* by microbiologist Joshua Lederberg. In his weekly column Lederberg, a professor at Stanford and a political liberal, discussed a wide range of scientific issues. On Sept. 30, 1967, he wrote that it would be "an interesting exercise in social science fiction" to "contemplate the changes in human affairs that might come about from the generation of a few identical twins of existing personalities" (Lederberg, "Unpredictable Variety Still Rules Human Reproduction," *Washington Post,* Sept. 30, 1967, p. A17). Such a casual attitude horrified Kass, who wrote an indignant letter to the editor. "It is unfortunate that Dr. Lederberg is either unaware or unwilling to discuss the moral and political problems involved [in human cloning]," he wrote; "it is shocking that he chooses to speak as if these questions are trivial." Kass's reaction to Lederberg's

column, and the way it changed his career, is described in Gina Kolata, *Clone: The Path to Dolly and the Road Ahead* (New York: William Morrow, 1998), p. 90.

"new holy war": Kass, "New Beginnings in Life," in Hamilton, *New Genetics and the Future of Man,* pp. 20, 21.

71 **"we cannot rightfully":** Paul Ramsey, *Fabricated Man* (New Haven: Yale University Press, 1970), p. 113.

he gave a summary: as described a year before the Kennedy Foundation meeting in Walter Sullivan, "Implant of Human Embryo Seems Near," *New York Times,* Oct. 29, 1970, p. 1.

72 **"In vitro fertilization":** The Kennedy Foundation never published the proceedings of the conference, but Ramsey published his own remarks in a two-part article, "Shall We 'Reproduce'?" *JAMA* 220, no. 10 (June 5, 1972): 1346–50 and no. 11 (June 12, 1972): 1480–85. This, and other quotes of Ramsey's from the Kennedy Foundation conference, are from that article. Details about the conference, including programs, letters of invitations, and a summary of the "Fabricated Babies" panel prepared for "Men and Molecules," are in the archives of the Joseph P. Kennedy, Jr., Foundation.

"prickly," "pontifical": Ronald A. Carson, "Paul Ramsey, Principled Protestant Casuist: A Retrospective," *Medical Humanities Review* 2, no. 1 (Jan. 1988): 26.

73 **Ramsey said he hoped:** in "Shall We 'Reproduce'?" (pt. 2). Ramsey repeated this "hope" in "Manufacturing Our Offspring: Weighing the Risks," *Hastings Center Report* 8, no. 5 (Oct. 1978): 7–9.

"It doesn't matter": Edwards and Steptoe, *Matter of Life,* p. 113.

"Good afternoon, ladies and gentlemen": transcript of "Men and Molecules."

74 **So much of medicine:** Edwards's recollections of the conference, including his interior monologue while under attack by Ramsey, are in Edwards and Steptoe, *Matter of Life,* p. 112.

"The whole edifice": *ibid.,* pp. 99–100.

rollicking, gossipy memoir: James Watson, *The Double Helix* (New York: W. W. Norton, 1968; reprint, 1980), p. 9.

75 **"Does this effective silence":** James Watson, "Moving Toward Clonal Man: Is This What We Want?" *Atlantic,* May 1971, pp. 50–53.

76 **"You can only go ahead":** Edwards and Steptoe, *Matter of Life,* p. 113.

"I accuse": *ibid.,* p. 114.

76 **"We shall do":** quoted in Stuart Auerbach, "Test-Tube Baby Study Spurs Ethical Dispute," *Washington Post,* Oct. 17, 1971, p. A1.

77 **likened to the Academy Awards:** Myra MacPherson, "Science-Spangled Awards Night," *Washington Post,* Oct. 18, 1971, p. B1.

5. Fits and Starts

78 **They began by:** P. C. Steptoe, R. G. Edwards, and J. M. Purdy, "Human Blastocysts Grown in Culture," *Nature* 229 (Jan. 8, 1971): 132–33.

79 **"Light, transparent":** Edwards and Steptoe, *Matter of Life,* pp. 94–95.

80 **"It is a terrible":** in Victor Cohn, "Scientists and Fetus Research: NIH Ruling May Halt What Some Deem Critical," *Washington Post,* Apr. 15, 1973, p. A1.

"If it is going to die": quoted in Victor Cohn, "Live-Fetus Research Debated: NIH Considering Ethics," *Washington Post,* Apr. 10, 1973, p. A1.

82 **"embryo wastage":** testimony of John D. Biggers before HEW's Ethics Advisory Board, Sept. 1978; see Henig, "In Vitro Fertilization," p. 685.

83 **The Tuskegee study:** for a brief summary of the experiment and its implications for biomedical ethics, see "Research Ethics: The Tuskegee Syphilis Study," online at members.home.com/jtstocks/prof/ethics/tuskg.html. Most of the details of the study reported here come from this article.

front-page story: Jean Heller, "Syphilis Victims in U.S. Study Went Untreated for 40 Years," *New York Times,* July 26, 1972, p. 1.

84 **DuVal asked:** Nancy Hicks, "Regulation Urged in Human Testing," *New York Times,* Mar. 21, 1973, p. 30.

"The withholding": quoted in UPI, "U.S. Syphilis Study Called 'Ethically Unjustified,'" *New York Times,* June 13, 1973.

85 **Ham's F-10:** Edwards and Steptoe, *Matter of Life,* p. 91.

"Chemicals and drugs": *ibid.,* p. 120.

86 **"the biggest threat":** quoted in "Artificial Fertilization," chap. 3 of Jones and Bodmer, *Our Future Inheritance,* p. 39.

6. Laboratory Ghouls

87 **took eight fetuses:** National Commission for the Protection of Research Subjects of Biomedical and Behavioral Research, *Research on*

the Fetus: Report and Recommendations (Washington, D.C.: Department of Health, Education, and Welfare, 1975) p. 14.

88 **"If there is a more unspeakable":** Victor Cohn, "NIH Vows Not to Fund Fetus Research," *Washington Post,* Apr. 13, 1973, p. A1.

"Why are they drawing": *ibid.*

89 **a panel of twenty-three:** the roster of names was read during Vande Wiele's testimony in the Del-Zio trial, transcript pp. 3166–67.

90 **"all experimentation":** Vande Wiele quoted the NIH policy in a letter to Shettles dated Sept. 14, 1973, which was an exhibit in the Del-Zio trial.

"To the best": Sadler's letter, dated Nov. 2, 1973, was admitted as a plaintiff's exhibit at the Del-Zio trial.

advisory commissions: The oversight mechanisms suggested in response to the Tuskegee revelations, which included a National Human Investigation Board to review all proposals for federal support of research involving human subjects, are described in Hicks, "Regulation Urged in Human Testing."

91 **As a compromise:** see Harold M. Schmeck, Jr., "Playing God: Necessary and Fearful," *New York Times,* July 15, 1973, p. 174.

The bill that was finally: see Culliton, "National Research Act," pp. 426–27.

7. Toward Happily Ever After

95 **He worried:** Vande Wiele described his state of mind on the morning of Sept. 13, as well as his actions, in testimony in the Del-Zio trial and in depositions beforehand. See trial transcript, pp. 3400–3407.

97 **"Dr. Allerand brought":** transcript of the tape-recorded conversation between Vande Wiele and Shettles, which was presented as documentation in the Del-Zio trial. All the quotes from this encounter are from that transcript, pp. 1–9.

99 **"Any man's":** Paul Ramsey, "Shall We 'Reproduce'? I. The Medical Ethics of In Vitro Fertilization," *JAMA* 220, no. 10 (June 5, 1972), p. 1347.

"all laws and rules": from the online journal The Gospel Herald, gospel-herald.com/genesis_studies/situational_ethics.htm.

100 **"dangerous ignorance":** John Fletcher, contribution to the roundtable discussion "In Vitro Fertilization," pp. 192–204. Elsewhere in the discussion he said, "We have to run some risks in embryo transfers, as in everything else, to know what we are doing. The refusal to run risk is sheer stasis, moral hardening of the arteries."

101 **"happy and hopeful":** Del-Zio trial transcript, p. 125.

"The procedure has": Sweeney's testimony, trial transcript, pp. 609 ff.

"I'm so sorry": Doris Del-Zio described her phone conversation with Shettles in her deposition in 1976.

102 **Toran-Allerand opened the door:** interview, March 2001. See also her testimony at the Del-Zio trial, transcript p. 2381.

103 **"To be right":** the Cajal quote was among Shettles's personal papers archived by Lana Shettles Callahan, shown to me on Nov. 5, 2002.

"certain irreconcilable": Shettles's resignation letter, to Presbyterian Hospital and the Columbia College of Physicians and Surgeons, dated Oct. 17, 1973 (though in a different letter Vande Wiele says the resignation was effective Oct. 15), was admitted as evidence in the Del-Zio trial.

Shettles continued: In a letter to Shettles on Dec. 12, 1973 (part of the court record in the Del-Zio trial), Vande Wiele said that "a number of people" had told him Shettles was seen "almost daily in different parts of the Medical Center," often wearing "scrub clothes and/or white coat." He asked Shettles to stay away. "Since you have no appointment either in the Medical School or the Hospital, I think this is misleading and I would like to advise you against this practice."

The first thing she did: interview in March 2001.

8. Baby Dreams

104 **"Dream that my little":** quoted by Maurice Hindle in the introduction to Mary Shelley, *Frankenstein* (reprint, New York: Penguin Books, 1992), p. xv. Details about the birth and death of the baby, named Clara, can be found online at www.english.udel.edu/swilson/mws/chrono.html. Mary Shelley had a great deal of death in her young life: Percy Bysshe Shelley drowned in a boating accident in the Gulf of Spezia in the summer of 1822, leaving behind the twenty-four-year-old Mary and their two-year-old son, Percy. Just two weeks before Shelley's death, Mary had had a miscarriage and had almost bled to death. Two more children had already died by that time: Clara, who had been given the same name as Mary's first baby, died of dysentery in Venice soon after her first birthday in 1818; William died of malaria at the age of three while living in Rome in 1819.

105 **In the dream:** Doris Del-Zio's testimony at trial, transcript p. 152.

"Battle of the Sexes": see Larry Schwartz, "Billie Jean Won for All Women," special to ESPN.com, espn.go.com/sportscentury/features/00016060.html.

106 **"As cardiology nurses":** Kristin Hall, who sued Northwestern, was eventually reinstated with back pay. For a full description of the case, see Robin Marantz Henig, "Nurse Power," *New Physician*, May 1976, pp. 28–34.

107 **only 7 percent:** the statistics are from the *Web Weekly* of Harvard Medical School: Sean Amos, "The Changing Face of Medicine," Nov. 19–26, 2001: webweekly.hms.harvard.edu/archive/2001/11_19/student_scene .html.

The Total Woman: see Sally Quinn, "The 'Total Woman' May Have Left the Feminists Behind," *Washington Post*, Jan. 27, 1978, p. D1.

108 **population was actually declining:** see the Web site of Zero Population Growth, now called the Population Connection, at www.population connection.org/About_Us/history.html.

7 percent of married couples: Ethics Advisory Board, *Report and Conclusions: HEW Support of Research Involving Human In Vitro Fertilization and Embryo Transfer* (Washington, D.C.: Department of Health, Education, and Welfare, May 4, 1979), p. 37. The authors calculated that about one-third of infertility cases were caused by a problem of the wife's, meaning that of 4.2 million infertile couples in the United States (7 percent of 60 million "reproductively active" couples), 1.4 million could trace the problem to the woman, and of those, 40 percent, or 560,000, could trace it specifically to diseased fallopian tubes.

109 **She had been:** John Del-Zio's testimony under direct examination, trial transcript, pp. 2325–27. Twenty-four years later, in an interview at his home in Fort Lauderdale on Nov. 20, 2002, John remembered this period differently, saying that Doris was not much sadder after the IVF attempt than she had been before. Keeping in mind that the Del-Zios may have exaggerated in court to make their $1.5 million case and that time may have dimmed John's recollection of Doris's mental state in the seventies, I have opted to accept the version they offered under oath five years after the event.

The Six Million Dollar Man: The show became so popular that its scripts were included in a pilot program to encourage schoolchildren in Washington, D.C. to read. Every week the networks would send the scripts of the latest installments of the year's most watched prime time shows — *Six Million Dollar Man, The Waltons, Little House on the Prairie, Welcome Back, Kotter, Sanford and Son, Rhoda, The Bionic*

Woman (a *Six Million Dollar Man* spinoff) — to public schools in the inner city to serve as a stimulus for lessons in vocabulary, plot structure, and reading. See Lawrence Feinberg, "Priming Time: TV Scripts Used to Help Pupils Learn to Read," *Washington Post,* Nov. 20, 1977, p. B1.

110 **"It's been five years":** plaintiff's exhibit in the Del-Zio trial.

111 **"couldn't look at my husband":** Doris Del-Zio's deposition, taken by James Furey in 1975, p. 56.

anachronistic police work: Barbara Culliton, "Grave-Robbing: The Charge Against Four from Boston City Hospital," *Science* 186 (Nov. 1, 1974): 420–23. The scientists had published their findings in *NEJM,* July 7, 1973; they found that pregnant women metabolized antibiotics differently from nonpregnant women and that clindamycin crossed the placenta more effectively than erythromycin.

112 **she was shopping:** Doris Del-Zio's trial testimony, transcript, p. 156.

"quite uncharacteristic": "The Great 'Test Tube' Baby Furor — So Far," *Medical World News,* Aug. 9, 1974, p. 15. Bevis's quotes and the quotes in response to his claims are from the same article.

114 **"bed of nettles":** ibid., p. 16. Sir John Peel, who made this comment, was the chairman of Bevis's scientific panel. The incident was also described in David M. Rorvik, "The Embryo Sweepstakes," *New York Times Magazine,* Sept. 15, 1974, p. 17.

on the recommendation: "Great 'Test Tube' Baby Furor," p. 16.

115 **"It would have been one thing":** Del-Zio trial transcript, p. 70.

At the urging: In an interview in November 2002, John Del-Zio remembered receiving phone calls from both Sweeney and Shettles urging him to sue Vande Wiele.

116 **"shocking, outrageous":** Judge Charles Stewart's Memorandum Decision, filed in U.S. District Court, Nov. 14 1978, p. 5.

typical settlement: James Furey, personal interview, spring 2001.

cardiac surgeons agreed: see Harry Schwartz, "A World Moratorium on Heart Transplants," *New York Times,* Aug. 22, 1971, p. E7.

9. Science on Hold

118 **"What do you want to do?":** quoted in Michael Rogers, "The Pandora's Box Congress," *Rolling Stone,* June 19, 1975.

119 **"We must remember":** Robert Sinsheimer, "Troubled Dawn for Genetic Engineering," *New Scientist* 68 (Oct. 16, 1975): 148.

wrote a letter: Paul Berg, et al., "Potential Biohazards of Recombinant DNA Molecules," *Science* 185, no. 303 (July 26, 1974), letters section.

120 **Vande Wiele seemed:** *International Conference on Biology and the Future of Man: Proceedings of a conference at the Sorbonne, Sept. 18–24, 1974* (Paris: University of Paris Press, 1976), pp. 400–403.

121 **"Debating the ethics":** Many of the details about the Asilomar conference, including all the quotes and jokes and the story about the butterflies, are from Rogers, "Pandora's Box Congress" and from his book based on the article, *Biohazard* (New York: Alfred A. Knopf, 1977).

122 **"The Berg experiment":** Nicholas Wade, "Microbiology: Hazardous Profession Faces New Uncertainties," *Science* 182 (Nov. 9, 1973): 566–67.

123 **"And if we cannot":** Sinsheimer, "Troubled Dawn."

124 **statement summarizing:** James D. Watson and John Tooze, *The DNA Story: A Documentary History of Gene Cloning* (San Francisco: W. H. Freeman, 1981), pp. 41–42.

overly tight controls: According to James Watson, one of the conveners of Asilomar, there would be a cost to crying wolf. "Inevitably," he later recalled, "those who wished to see recombinant DNA research progress quickly found the proposed guidelines much too restrictive; those who wished a virtual halt to the research found them too lax." Watson and Tooze, *DNA Story,* p. 66.

125 **"Have there been":** Rogers, *Biohazard,* pp. 161–62.

126 **"misunderstood the whole":** quoted in *Biohazard,* p. 162.

128 **The President's Commission:** At its first meeting in December 1974, the group was instructed to come up with a report on fetal research by May 1, according to Victor Cohn, "Much Research Blocked by Fetal Law," *Washington Post,* Jan. 11, 1975, p. A8.

Albert Jonsen: his background is posted on the Web site of the University of Washington, where he is an emeritus professor in the department of medical history and ethics, which he headed from 1987 to 1999. See depts.washington.edu/mhedept/facres/aj_bio.html.

They heard oral testimony: the list of people testifying before the commission, and summaries of their testimony, appear in the commission's final report, *Research on the Fetus,* pp. 41–51. The testimony presented at the February and March hearings, including that of Monsignor McHugh, is summarized on p. 74.

129 **"no intrusion":** *ibid.*, p. 74.

130 **recommendations had to wend:** the process is described in Culliton, "National Research Act," p. 427.

Soupart had asked: his proposal is described in several places, including Kass, "'Making Babies' Revisited."

131 **Weinberger lifted the ban:** see Cohn, "Fetal Research Ban Lifted."

10. The First One

133 **"I had never":** John and Lesley Brown with Sue Freeman, *Our Miracle Called Louise* (London: Paddington Press, 1979), p. 84; *"Go on, go":* p. 84; *"It just didn't occur":* pp. 157–58.

135 **"Does the perfection":** in Grobstein, *From Chance to Purpose*, pp. 155–56. (Appendix I)

Some people urged: summaries of testimony at the various hearings are in Ethics Advisory Board, *HEW Support of Research Involving Human In Vitro Fertilization and Embryo Transfer: Report and Conclusions* (Washington, D.C.: Department of Health, Education, and Welfare, May 4, 1979).

136 **"The essence":** quoted in Henig, "In Vitro Fertilization," pp. 685–86.

137 **"the humanness":** Kass's comments to the board are in Kass, "'Making Babies' Revisited," *Public Interest*, Winter 1979.

138 **"of uncertain or remote risk":** Ethics Advisory Board, *HEW Support*, p. 102.

139 **Jones received a call:** Howard Jones told me of Mason Andrews's invitation and the couple's decision in an interview in October 2001.

140 **"the yokel":** H. L. Mencken, *A Mencken Chrestomathy* (New York: Alfred A. Knopf, 1949), p. 8. The quote originally appeared in *The Smart Set*, Mencken's column in the *Baltimore Sun*, in June 1920.

141 **"I'm the first one":** *Our Miracle*, p. 158.

a reported £300,000: The exact amount is a matter of some dispute. John Brown, in *Our Miracle*, put the amount at more than £20,000 but less than £100,000. The £300,000 figure comes from Victor Cohn, "Test-Tube Baby Reported Near," *Washington Post*, July 12, 1978, p. A1.

John's first paypacket: *Our Miracle*, p. 73.

11. A Baby Clone

143 **"The test tube":** quoted in Edwards and Steptoe, *Matter of Life,* pp. 84–85.

144 **Rorvik wrote about:** *In His Image: The Cloning of a Man* (Philadelphia: J. B. Lippincott, 1978), pp. 19–23, 64–65.

"virtually the only": Rorvik, "Embryo Sweepstakes"; *"daring experiments":* "Test-Tube Baby is Coming"; *"contributing substantially":* ibid.

145 **"After all":** Shettles mentioned Rorvik's phone call and his interest in cloning Max to Angela Robinson, a reporter for *Newsday,* who tracked him down at the Vermont house he was renting — the first real house he had lived in, complete with white picket fence, in his entire adult life — while he worked at a small country hospital. See "MD Says He'd Clone on Request," *Newsday,* Mar. 5, 1978.

146 **dozens of attempts:** Rorvik was not specific about how many attempts Darwin made or how many times he tried to insert a fertilized egg into a surrogate mother's uterus. However, when Rorvik mentioned that it might take as many as one hundred attempts before he succeeded, Darwin thought that number was much too high. Rorvik, *In His Image,* p. 177.

Some observers: Negative reviews of *In His Image* included Jerry Bishop, "Implausible Tale of Genetic Tinkering," *Wall Street Journal,* Mar. 29, 1978, p. 22; and Michael Crichton, "Cloning Around," *New York Times Book Review,* Apr. 23, 1978, p. 2.

armchair analyses: The comments by Mintz and Jaroff are from Peter Gwynne, "All About Clones," *Newsweek,* Mar. 20, 1978, p. 68; *"his desire to write":* Herbert Mitgang, "Cloning Becomes a Publishers' Experiment," *New York Times,* Mar. 11, 1978, p. 26.

147 **"After one scientist":** joke number 818 on the Humor Mall Web site, www.humormall.com, which is posted weekly. It was originally published in 1978 in a biweekly humor service called "comedy/update," as were the IRS and Erector set jokes.

148 **something akin to cloning:** Hoppe and Illmensee's experiment is described in Kolata, *Clone,* pp. 126–27. It is also described, along with the testimony from the other scientists appearing before the Rogers subcommittee, in Victor Cohn, "Research Shows Human Cloning Doubtful," *Washington Post,* June 1, 1978, p. A6.

149 **One hundred thousand copies:** Constance Holden, "British Scientist

Sues over Clone Book," *Science* 201 (July 28, 1978): 326. *Rorvik reportedly earned:* Philip J. Hilts, "Publisher Settles in Clone Book Case," *Washington Post,* Apr. 8, 1982, p. A5. One of the sources mentioned in Rorvik's book, J. Derek Bromhall of Oxford, filed suit against the author and his publisher, J. B. Lippincott, and the amount of Rorvik's earnings was revealed in the settlement. According to Hilts, Lippincott earned $730,000 from the book before expenses and agreed to pay Bromhall $100,000.

12. Hang On

150 **The first sign:** Brown's pregnancy history is recorded in Edwards and Steptoe, *Matter of Life,* p. 165.

151 **some 40 percent:** Dianne Hales, and Robert K. Creasy, *New Hope for Problem Pregnancies* (New York: Harper & Row, 1982), p. 4. *Toxemia occurred:* p. 106. Toxemia, later called preeclampsia, is most likely to occur in women younger than eighteen or older than thirty-five.

"What if you were": Elliott Phillip, a London gynecologist, recalled his conversation with Steptoe in the BBC production "The Baby Makers," (2001).

she burst into tears: Edwards and Steptoe, *Matter of Life,* p. 166.

152 **said to be:** *Our Miracle,* p. 162.

13. Fooling Mother Nature

155 **reporters jostled:** Bower described the scene in a conversation in the judge's robing room; see trial transcript, p. 8.

156 **"It's not nice":** see "70s TV for Grown-Ups" Web site, www.rt66.com/dthomas/70s/adulttv/adulttv.html.

157 **dangerous near-misses:** The Rocky Flats accident took place on May 11, 1969, a precursor to an accidental release of radiation at the Hanford nuclear weapons plant in 1974 and to a near meltdown at the Three Mile Island nuclear power plant in 1979. See www.multiline.com.au/~mg/legion1.html.

158 **The original date:** the trial delays were outlined by plaintiff's attorney Michael Dennis, trial transcript, pp. 10–11. Dennis said the trial date

was first set for December 1977, but in the judge's memorandum decision to deny a motion to dismiss after the verdict was in, the first date was given as Nov. 10, 1977.

Bower was a brusque: Bower's demeanor and courtroom behavior were described by Michael Dennis in an interview at his home in Naples, Florida, January 2002, and also by James Furey in an interview at his office in Hempstead, New York, May 2001.

159 **"Are we to be denied":** trial transcript, p. 13.

empaneled a jury: the description of the jurors is from a letter describing the trial in great detail, written by James Furey to the Hartford Insurance Group, to the attention of Eugene T. DeCanio, on Aug. 31, 1978. The *Times* reporter was the late Joan Cook.

160 **the young man headed:** Shettles's testimony, trial transcript p. 1464. Shettles said the young man who transported the gametes was Sweeney's son, but Sweeney testified that it was his stepson: trial transcript, pp. 580–81.

161 **"After you read":** trial transcript, p. 380.

163 **"Those two specimens":** *ibid.*, pp. 365–66.

164 **"please don't brainwash me":** *ibid.*, p. 364. See also D. J. Saunders and Paul Meskil, "Doris Pleads: Don't Brainwash Me," *New York Daily News*, July 20, 1978, p. 3.

he asked her to agree: trial transcript, pp. 260–75.

166 **TEST-TUBE MA'S SPIRIT:** article by D. J. Saunders and Marcia Kramer, *New York Daily News*, July 21, 1978, p. 5.

"I don't see": discussion in the robing room, July 21, 1978, trial transcript, pp. 776–77.

167 **"In other words":** trial transcript, p. 189.

call him a leprechaun: phone interview with Doris Del-Zio, March 2001.

when Sweeney applied: Vande Wiele testified that he took Sweeney to dinner to let him "plead his case" about why he should get the St. Luke's appointment and that Vande Wiele initially supported him, but that after the selection committee found out about Sweeney's book, *Woman's Doctor*, "there was no further discussion about his qualifications," trial transcript, pp. 3374–75.

168 **"Are you still":** Vande Wiele's deposition, Dec. 22, 1975.

"What the hell": the quotes are from William J. Sweeney III with Barbara Lang Stern, *Woman's Doctor: A Year in the Life of an Obstetri-*

cian-Gynecologist (New York: William Morrow, 1973), pp. 51, 68, and 75.

169 **"I did not consider":** trial transcript, p. 614. Sweeney's testimony was also described in Judith Cummings, *New York Times,* July 25, 1978, sec. 2, p. 3.

14. Pandora's Baby

170 **Reporters from around:** from Edwards and Steptoe, *Matter of Life,* pp. 164–65. Most of the descriptions of this night's events are from this book as well: *Steptoe made plans,* p. 172; *So he slipped* away, p. 173; *earned pocket money,* p. 65; *fighting about film rights,* p. 174.

15. Normality

173 **IT'S A GIRL:** *New York Daily News,* July 26, 1978. The full headline, which took up the entire front page, was 1ST TEST TUBE BABY IS BORN —IT'S A GIRL; CONDITION "EXCELLENT." The *Daily Mail* headline was "AND HERE SHE IS . . . THE LOVELY LOUISE," with a huge photo of Louise swaddled in a white blanket, her hand brought daintily up to her lips.

"The normality": quoted in Victor Cohn, "U.S. Scientists Urge More Study Before Test-Tube Babies," *Washington Post,* July 27, 1978, p. A19.

174 **three major religions:** The priest, minister, and rabbi are quoted in "Religion vs. Science: Medical Miracle a Thorny Issue for Theologians," *New York Daily News,* July 30, 1978, p. 5.

This was the pattern: The committee's report, though written in 1973, was not published until 1977. It pointed out that several similar reports helped establish the field of technology assessment. Two were commissioned by the U.S. House of Representatives' Committee on Science and Astronautics and delivered to the committee by the National Academy of Sciences in July 1969: *Technology: Processes of Assessment and Choice* and *A Study of Technology Assessment.*

175 **"The initial reactions":** National Academy of Sciences, *Assessing Biomedical Technologies* (Washington, D.C.: National Academies Press, 1977), p. 19. In a personal interview in June 2002, Leon Kass said that

the report was "so controversial" that the academy "found it dangerous" and resisted releasing it until the National Science Foundation said that it would publish the report itself if the academy refused to. The report dealt with technology assessment in four areas: sex selection, aging research, behavioral research, and in vitro fertilization.

176 **"Can a test tube baby":** Howard Jones has related the story of how he got involved in IVF several times, including in his keynote address at the symposium "Optimizing ART: A Symposium to Honor Howard and Georgeanna Jones," Williamsburg, Virginia, Apr. 4–7, 2002.

whom she named Georgia: interview with Howard Jones, October 2001.

16. Prometheus Unbound

178 **"We do not concede":** closing arguments, trial transcript, pp. 4198 ff.

Under questioning: *ibid.*, pp. 3523 ff.

179 **"The article is entitled":** Landrum B. Shettles, "The Morula Stage of Human Ovum Developed in Vitro," *Journal of Fertility and Sterility* 6 (July/Aug. 1955): 287.

180 **Glancing out the window:** Bower described his observation of Shettles in the robing room, trial transcript, pp. 1532–33.

181 **"Each egg would have been":** *ibid.*, p. 1567.

Furey asked repeatedly: *ibid.*, pp. 1614–15.

183 **caught off guard:** Bower said the reporter had "sidled" up to him, ibid. p. 506. *Stewart was furious: ibid.*, p. 509.

184 **His credentials:** *ibid.*, p. 3126.

185 **"The first reason":** Vande Wiele enumerated his reasons at several points in his testimony; see, for example, transcript, p. 3170.

186 **"I am not a coward":** *ibid.*, pp. 3390–91.

In private he referred: in a personal interview in January 2002, Dennis kept calling Vande Wiele "Wilhelm," almost as though he had forgotten that his first name was really Raymond.

187 **"Is it your view":** trial transcript, p. 3542.

188 **two staff members:** Henig, "In Vitro Fertilization," p. 685.

189 **pretty wild speculation:** *ibid.*, and Edwards and Steptoe, *Matter of Life,* p. 90.

"They're having the biggest": jokes originally published in August 1978 in the biweekly humor service "comedy/update."

190 **98 percent:** Gallup poll cited in Ethics Advisory Board report, May 4, 1979, p. 88.

191 **According to her testimony:** trial transcript, pp. 2816–17, 2824, 2825, and 2979. She testified right before Raymond Vande Wiele, although their testimony is presented here in reverse order. Vande Wiele testified on days eighteen through twenty of the trial (Aug. 9, 10, and 11), and Jagiello testified on Aug. 7, 8, and 9.

17. Verdict

194 **"greater weight":** charge to the jury, Del-Zio trial documents. The notes sent out by the jury through the court bailiff are in the same document boxes.

196 **Nervous and edgy:** as described in Doris Del-Zio with Suzanne Wilding, "I Was Cheated of My Test-Tube Baby," *Good Housekeeping*, Mar. 1979, pp. 135 ff.

One woman on the jury: phone interview with Hugh Lawless in March 2001.

At 7:30 p.m.: Furey to Hartford Insurance Group, p. 13.

197 **The jury foreman:** Hugh Lawless, phone interview, March 2001.

"atrocious" behavior: Del-Zio trial documents. See also *New York Times*, Aug. 19, 1978.

18. Right to Life

201 **Overall, 60 percent:** Ethics Advisory Board, *HEW Support of Research*, May 4, 1979, p. 88.

202 **two failed pregnancies:** mentioned in several sources, including M. J. Ashwood-Smith, "Robert Edwards at 55," *Reproductive Biomedicine Online* 4, supp. 1 (Mar. 2002): 3.

a triploidy: Edwards emphatically denied the reports of the sixty-nine-chromosome fetus in *Human Conception in Vitro: Proceedings of the First Bourn Hall Conference* (New York: Academic Press, 1982), p. 344. But the claim continued to be made; see, for example, Andrews, *Clone Age*, p. 19.

"artificial conditions": Edwards and Steptoe, *Matter of Life*, p. 104.

203 **in India:** the Indian birth was widely reported at the time, but the doc-

tors involved never documented it and never published details of the birth in a medical journal, ultimately earning the skepticism of the scientific community.

204 **"Mason Delivers":** Sandra G. Boodman and Dale Russakoff, "'It's Positive' — How Norfolk School Created Test-Tube Pregnancy," *Washington Post,* May 17, 1981. Andrews served on the Norfolk City Council from 1974 to 2000, including two years as mayor (1992–1994), and was named Norfolk Citizen of the Year in 1968. The article described his tie clasp as being in the shape of a hound dog, but Howard Jones recalls it as a llama.

exclusive retreats: Georgeanna Jones invited Robert Edwards to be the keynote speaker at the American Fertility Society meeting in 1971, the year she was the group's president.

205 **certain conditions:** Victor Cohn, "Ethics Board Gives Backing To Test-Tube Baby Research," *Washington Post,* Mar. 17, 1979, p. A1; see also Robin Marantz Henig, "Go Forth and Multiply, Ethics Board Tells Scientists," *BioScience* 29, no. 5 (May 1979): 321–23.

206 **"imperfect babies":** Glenn Frankel, "Test-Tube Baby Clinic," *Washington Post,* Jan. 8, 1980, p. C1.

"very exciting": The federation's executive director, Rev. N. J. L'Heureux Jr., in "Religion vs. Science: Medical Miracle a Thorny Issue for Theologians," *Daily News,* July 30, 1978, p. 5.

208 **"The last thing":** Frankel, "Test-Tube Baby Clinic."

In the fifth century B.C.: see Halloween-on-the-Web (wilstar.com/holidays/hallown.htm); and Charles Panati, *Extraordinary Origins of Everyday Things,* (New York: Harper Collins, 1987). I obtained some information about the history of Halloween from Joseph Gahagan, University of Wisconsin-Milwaukee, personal correspondence, 1997.

209 **This time, however:** details about the Oct. 31, 1979, certificate of need hearing are from Howard Jones, interview, October 2001.

health commission voted: Susan Lockamy, "In-Vitro Proposal Endorsed by Panel," *Virginian-Pilot,* Nov. 14, 1979.

Senator Orrin Hatch: Sandy Baksys, "US Senator Seeks Hearings on In Vitro Labs," *Ledger-Star,* Jan. 9, 1980.

threat of a lawsuit: Susan Lockamy, "Injunction Sought on Baby Lab," *Virginian-Pilot,* Feb. 11, 1980.

210 **"My investigations":** Lawrence D. Pratt, "Safeguards Needed for In-Vitro Experiments," *Virginian-Pilot,* Mar. 10, 1980, p. A11. Only one of the two IVF pregnancies Pratt described was abnormal, the one that was a

triploidy fetus; the other one miscarried, possibly because the mother overexerted herself while hiking.

211 **the senators asked:** Sandy Baksys, "Senators Ask for Review of HEW's In Vitro Study," *Ledger-Star,* Feb. 22, 1980.

Abram declined: AP, "Further In Vitro Studies Opposed," *Ledger-Star,* May 17, 1980.

"shake the foundation" unsigned editorial, "Test Tube Babies," *Washington Post,* Jan. 19, 1980, p. A14.

212 **"If one reviews":** Landrum Shettles, "In Vitro Report Questioned," *American Medical News,* Aug. 15, 1980, p. 5.

213 **Estes operation:** see "IVF — Historical Perspective," in Fishel, *In Vitro Fertilization,* pp. 6–7.

"Why the big push": Charles R. Dean, "In-Vitro: Fraud or Fertility Assistance?" *Virginian-Pilot,* Nov. 14, 1980.

214 **Once someone mailed:** The Joneses' setbacks are described in Betty Cuniberti, "For the Test-Tube Doctors, It Was a Triumph of Will," *Los Angeles Times,* Jan. 5, 1982, p. 1.

215 **Two molecular biologists:** Lee M. Silver, *Remaking Eden: How Genetic Engineering and Cloning Will Transform the American Family* (New York: Avon Books, 1997), p. 270.

216 **"We just wanted to know":** in Brendan A. Maher, "Test Tubes with Tail: How Mice Help Play Out Science's Best Laid Plans," *Scientist* 16, no. 3 (Feb. 4, 2002): 22.

19. Opening Pandora's Box

217 **a capital of culture:** see Travel Online Melbourne Web site, www.melbourne.visitorsbureau.com.au/.

218 **first test tube twins:** Stephen and Amanda Mays were born in Melbourne on June 6, 1981.

"There was tremendous excitement": quoted in "Australian Story," television show that aired in Australia in 1999; see transcript at www.abc.net.au/austory/archives.

Scientists at the two clinics: according to Bill Gibbons, in an interview in October 2001 at his office in Norfolk.

the Joneses implanted: Suzanne Viau, "Timing Is Clinic's Major Obstacle," *US Medicine,* Oct. 15, 1980, p. 1.

"The embryo will wait": interview with Howard Jones in January 2002. The same saying was quoted by John D. Biggers, professor of physiology at Harvard, during his September 1978 testimony before the Ethics Advisory Board; see in Henig, "In Vitro Fertilization," p. 685.

219 **"We really don't know":** *ibid.*

220 **Boston obstetrician was convicted:** Kenneth Edelin, the chief resident at Boston City Hospital, received a suspended sentence and never went to jail, becoming instead a luminary in the pro-choice movement. In 1981 a play about his trial opened off Broadway; according to the *New York Times* theater critic Frank Rich, it was "less well-suited to the stage than to a brief magazine article." (Frank Rich, "A Drama in Court: 'The Meaning of Words' " *New York Times,* June 3, 1981, p. C22.)

Surrounded all day: Bella English, "The Test-Tube Girl Next Door," *Boston Globe,* Jan. 21, 1999.

221 **Judy talked about:** interview with Judy Carr at her home in Massachusetts in February 2002.

222 **"Well, her temperature":** interview with Howard Jones in October 2001 and a follow-up phone conversation in March 2003.

223 **"All we can do":** *ibid.*

Pierre Soupart: According to the obituary in the *New York Times* (June 12, 1981, p. D15), Soupart's life history was similar to Raymond Vande Wiele's in several ways: each man was born and raised in Belgium, was a professor of obstetrics and gynecology, and was the father of three girls.

224 **"People expected her":** quoted in *The Baby Maker,* Australian documentary film about Carl Wood, produced by Brigid Donovan, and broadcast Oct. 5, 2001.

225 **thermostat was set:** English, "Test-Tube Girl Next Door."

his own seventy-first: a brief biography of Jones can be found at www.inciid.org/members/howard_jones.html.

226 **That distinction belonged:** Bonnie Remsberg, "It's a Girl," *Ladies Home Journal,* Jan. 1982, p. 75 ff.

An estimated one hundred: Walter Sullivan, " 'Test-Tube' Baby Born in U.S., Joining Successes Around the World," *New York Times,* Dec. 29, 1981, p. C1.

227 **"Everybody is anti-abortion":** Cuniberti, "For the Test-Tube Doctors, It Was a Triumph of Will."

227 **$5.5 million lawsuit:** "Test Tube Baby Doctor Sues for Libel," *Arlington Journal,* May 18, 1982.

"There is no more demanding": quoted in Lisa Ellis, "Businesses Name In-vitro Lab Pioneer Woman of the Year," *Norfolk Virginian-Pilot,* May 29, 1982, p. B1.

228 **"making yesterday's miracle":** Victor Cohn, "Norfolk Team in Forefront of Test-Tube Baby Boom," *Washington Post,* Sept. 13, 1982, p. A2.

20. Tables Turned

229 **for Michelangelo to paint:** According to the Sistine Chapel Web site, www.kfki.hu/~arthp/tours/sistina/index_a.html, Pope Julius II della Rovere commissioned Michelangelo in 1508 to repaint the ceiling of the chapel, which had been built between 1475 and 1483 and originally painted with frescoes in 1481–82. Michelangelo worked on the ceiling from 1508 to 1512 and then from 1535 to 1541.

230 **"eighteen, nineteen, or twenty":** and other quotes, Judy Klemesrud, "Fertilization Clinic Open in New York," *New York Times,* Jan. 17, 1983, p. B6.

231 **laws limiting or prohibiting:** Otto Friedrich, "'A Legal, Moral, Social Nightmare,'" *Time,* Sept. 10, 1984, p. 54.

232 **Luigi Mastroianni:** e-mail message from Mastroianni to me on Feb. 20, 2002.

On Saturday he went: Michel Ferin said in an interview that Vande Wiele saw a slide presentation on the last day of his life, but Raphael Jewelewicz, another Columbia colleague and longtime collaborator, said in an interview in May 2001 that Vande Wiele saw a Woody Allen movie that night. I imagine that he did both.

Vande Wiele woke up: from interview with Seymour Lieberman, May 2001.

21. From Monstrous to Mundane

233 **In 1998 nearly 82,000:** "Use of Assisted Reproductive Technology—United States, 1996 and 1998," *Morbidity and Mortality Weekly Report* 51, no. 5 (Feb. 8, 2002): 97–101. The report is available at the CDC Web site: www.cdc.gov/mmwr/preview/mmwrhtml/mm5105a2.htm.

An estimated 450: Joyce Zeitz, American Society of Reproductive Medicine, e-mail message, Apr. 2003.

235 **ovarian hyperstimulation syndrome:** the rate of 1 to 10 percent is claimed in A. D'Angelo and N. Amso, "'Coasting' (Withholding Gonadotrophins) for Preventing Ovarian Hyperstimulation Syndrome," *Cochrane Library* no. 4 (2002). But according to the Web site www.pregnancymd.org/ovarian-hyperstimulation-syndrome-ohss.htm, the rate is between 1 and 2 percent.

236 **Twins are twice as likely:** Laura A. Schieve et al., "Live-Birth Rates and Multiple-Birth Risk Using In Vitro Fertilization," *JAMA*, 282, no. 19 (Nov. 17, 1999): 1832–38.

The world's first IVF quadruplets: UPI, "World's First Test Tube Quads Born," *Charlottesville (Va.) Daily Progress*, Jan. 6, 1984; AP, "First 'Test Tube' Quadruplets," *Strasburg Northern Virginia Daily*, Jan. 7, 1984.

The real news: "Test-tube Quintuplets, First in U.S., Doing Well," *St. Petersburg Times*, Jan. 13, 1988, p. 1A.

success rates: The 1992 Fertility Clinic Success Rate and Certification Act requires all U.S. clinics performing ART to report data annually to the U.S. Centers for Disease Control for every ART procedure initiated, including IVF, GIFT, and ZIFT. Each procedure is classified according to whether the eggs were the patient's own or were donated by another woman, whether the embryos transferred were freshly fertilized or previously frozen, and whether the embryos were transferred into a gestational surrogate mother rather than the patient herself. Data from 1996, when clinics first began submitting their data to CDC through the Society for Assisted Reproductive Technology reporting system, as well as from 1998, the most recent year for which statistics are available, are in CDC, *Morbidity and Mortality Weekly Report* 51, no. 5 (Feb. 8, 2002): 97–101.

237 **In Europe:** Comparisons of U.S. and European rates of IVF multiples are from Patricia Katz, Robert Nachtigall, and Jonathan Showstack, "The Economic Impact of the Assisted Reproductive Technologies," *Nature Medicine* 8, no. 51, *Fertility Supplement* (2002): S30.

238 **Small-scale studies:** Robert Lee Hotz, *Designs on Life: Exploring the New Frontiers of Human Fertility* (New York: Pocket Books, 1991), pp. 22–23. The medical risks of IVF are also addressed in Robert M. L. Winston and Kate Hardy, "Are We Ignoring Potential Dangers of In Vitro Fertilization and Related Treatments?" *Nature Medicine* 8, no. 51, *Fertility Supplement* (2002): S14–S18.

239 **two unrelated studies:** Laura A. Schieve et al., "Low and Very Low Birth Weight in Infants Conceived with Use of Assisted Reproductive Technology," *NEJM* 346, no. 10 (Mar. 7, 2002): 731–37 and Michèle Hansen et al., "The Risk of Major Birth Defects after Intracytoplasmic Sperm Injection and in Vitro Fertilization," *ibid.*, pp. 725–30.

240 **reports emerged of higher rates:** Michael R. Debaun, Emily L. Niemitz, and Andrew P. Feinberg, "Association of In Vitro Fertilization with Beckwith-Wiedemann Syndrome and Epigenetic Alteration of LIT1 and H19," *American Journal of Human Genetics* 72 (2003): 156–60, in which the authors conclude that "at this point we do not know what specific steps of ART are responsible for the association with BWS. Potential factors could include culture conditions for the ovum, length of exposure to specific media or growth factors within it, the stage of differentiation of sperm at the time of ICSI, or ICSI itself."

241 **bladder exstrophy:** Hadley M. Wood, Bruce J. Trock, and John P. Gearhart, "In Vitro Fertilization and the Cloacal-Bladder Exstrophy-Epispadias Complex: Is There an Association?" *Journal of Urology* 169 (Apr. 2003): 1512–15.

The first frozen embryo birth: Andrews, *Clone Age,* p. 67; and David Perlman, "Brave New Babies," *San Francisco Chronicle,* Mar. 5, 1990, p. B3. The current number of frozen embryos in storage was provided by the American Society of Reproductive Medicine in July 2003.

number of frozen embryos: Nicholas Wade, "Clinics Hold More Embryos Than Had Been Thought," *New York Times,* May 9, 2003, p. A24.

242 **"an emotional boon":** Andrea L. Bonnicksen, *In Vitro Fertilization: Building Policy from Laboratories to Legislatures* (New York: Columbia University Press, 1989), p. 31.

"Basic functions": Lord Winston to a reporter for the *Independent,* quoted in "IVF Safety to Be Examined," *Human Genetics Alert,* Oct. 22, 2002. Winston's comments were posted on that date at news.bbc .co.uk.

22. Pandora's Clone

245 **"surreal":** AP, "Experts Struggle with Cloning Claim" *New York Times* Web site, Dec. 27, 2002: www.nytimes.com/aponline/national/AP-Cloning-Reax.html.

Valiant Ventures Ltd.: Virginia Morell, "A Clone of One's Own," *Discover* 19, no. 5 (May 1998): 84.

The clone was a healthy: The announcement and reactions to it are from Donald G. McNeil, Jr., "Religious Sect Announces the First Cloned Baby," *New York Times,* Dec. 27, 2002, or www.nytimes.com/2002/12/27/health/27CND-CLON.html.

246 **Their true goal:** Clifford Krauss, "Earthlings, the Prophet of Clone Is Alive in Quebec," *New York Times,* Feb. 24, 2003, p. A4.

248 **Time/CNN news poll:** Michael Shermer, "I, Clone," *Scientific American,* Apr. 2003.

249 **"Research cloning":** Charles Krauthammer, "A Secular Argument Against Research Cloning," *New Republic,* Apr. 29, 2002. See www.tnr.com/doc.mhtml?i=20020429&s=krauthammer042902&c=1. All of the Krauthammer quotes are from that article.

251 **"She had taken":** Richard Eder, "A Gene off the Old Block: A Clone's Identity Crisis" (book review), *New York Times,* Dec. 19, 2002, p. E15.

252 **Dolly the sheep died:** see, for example, Jim Gilchrist, "A Sheep of Faith," *New Scotsman,* Feb. 19, 2003, news.scotsman.com/index.cfm?id=208722003. Gilchrist mentions large-fetus syndrome as one complication of cloning. Another article, "Study Highlights Threat of Incomplete Genes to Cloned Humans," posted on the *ABC News* Web site (abcnews.com) on May 30, 2002, refers to problems of gene expression, mentioning that in one study of a group of cloned cows, nine out of ten genes located on the X chromosome were abnormal. And Rick Weiss, "Clone Defects Point to Need for 2 Genetic Parents," *Washington Post,* May 10, 1999, p. A1, mentions other problems associated with clonal pregnancies: placental abnormalities, edema, three to four times the normal rate of umbilical cord problems, and severe immunological deficiencies.

255 **"There are no new":** quoted in Howard Markel, "A Conversation with Harold Shapiro: Weighing Medical Ethics For Many Years to Come," *New York Times,* July 2, 2002, p. F6.

256 **He also posted literature:** www.bioethics.gov.

257 **"current and projected practices":** President's Council on Bioethics, *Human Cloning and Human Dignity: An Ethical Inquiry* (Washington, D.C.: July 2002), is available online at www.bioethics.gov.

258 **"the E. coli":** Marc Wortman, "The Mouse That Roared," *Yale Medicine* (fall/winter 1999–2000).

23. Mixed Blessings

262 **"test tube big girl":** *Sunday Mirror,* July 24, 1988, quoted in Robert Edwards, *Life Before Birth: Reflections on the Embryo Debate* (New York: Basic Books, 1989), p. 21. At Louise's twentieth birthday party, Tessa Jowell was the health minister in attendance, and Sally Keeble was the MP from Northampton who had had an IVF baby. The event was hosted by BBC reporter Cheryl Baker.

Natalie announced: Rachael Bletchly and David Newman, "She's My Double Miracle," *Mirror,* July 25, 1999, p. 4. Natalie's pregnancy and delivery were reported online at organon-conferences.com/eshre2002 news/product_information/fert_gen_info3.html.

263 **William Sweeney:** Wolfgang Saxon, "Dr. William Sweeney 3d, Researcher on In-Vitro Fertilization" (obituary), *New York Times,* Nov. 1, 1997, p. D16.

264 ***Glamour* magazine:** Elizabeth Carr's "as told to" article by Susan Dominus, "Why I Support Stem Cell Research," appeared in November 2001.

265 **"Who wants an ugly baby?":** Carolyn Abraham, "Gene Pioneer Urges Dream of Human Perfection," *Toronto Globe and Mail,* Oct. 26, 2002. The article was reprinted on the Web site of the Gairdner Foundation of Canada (www.gairdner.org/news50.html), a nonprofit organization that recognizes achievement in biomedical research, which gave Watson a lifetime achievement award in October 2002.

Selected Readings

Africano, Lillian. "I Was a Test Tube Baby." *Woman's Day,* July 26, 1977, p. 36 ff.

Andrews, Lori B. *The Clone Age: Adventures in the New World of Reproductive Technology.* New York: Henry Holt, 1999.

Arditti, Rita, Renata Duelli Klein, and Shelley Minden, eds. *Test Tube Women: What Future for Motherhood?* London: Pandora Press, 1984.

Barthel, Joan. "'Just a Normal, Naughty Three-Year-Old.'" *McCall's,* May 1982, p. 78 ff.

Bataille, Janet. "Research in Human Embryos Raises Fear and Hope," *New York Times,* Mar. 3, 1980, p. A14.

Berg, Paul et al. "Potential Biohazards of Recombinant DNA Molecules" (letter). *Science* 185, no. 303 (July 26, 1974).

Bonnicksen, Andrea L. *In Vitro Fertilization: Building Policy from Laboratories to Legislatures.* New York: Columbia University Press, 1989.

Brown, John, and Lesley Brown, with Sue Freeman. *Our Miracle Called Louise.* London: Paddington Press, 1979.

Clark, LeMon, and Isadore Rubin. "Dare I Have a Test Tube Baby?" *Sexology* 28, no. 6 (1961):364–69.

"Conceiving the Inconceivable" (editorial). *New York Times,* July 28, 1978, p. A22.

"Conception in a Watch Glass" (editorial). *New England Journal of Medicine* 217, no. 17 (Oct. 21, 1937): 678.

Edwards, R. G. "Fertilization of Human Eggs in Vitro: Morals, Ethics and the Law." *Quarterly Review of Biology* 49 (Mar. 1974): 3–26.

———. *Life Before Birth: Reflections on the Embryo Debate.* New York: Basic Books, 1989.

———. "Test-tube Babies, 1981." *Nature* 293 (Sept. 24, 1981): 253–56.

Edwards, Robert, and Patrick Steptoe. *A Matter of Life.* New York: William Morrow, 1980.

Etzioni, Amitai. *Genetic Fix: The Next Technological Revolution.* New York: Harper & Row, 1973.

———. "The 'Slippery Slope' of Science." *Science* 183 (Mar. 15, 1974): 1041.

Fishel, Simon, ed. *In Vitro Fertilization: Past, Present, and Future.* Washington, D.C.: IRL Press, 1988.

Francoeur, Robert T. *Utopian Motherhood: New Trends in Human Reproduction.* Garden City, N.Y.: Doubleday, 1970.

Gaylin, Willard. "The Frankenstein Myth Becomes a Reality." *New York Times Magazine,* Mar. 5, 1972, p. 12 ff.

"The Great 'Test Tube' Baby Furor — So Far." *Medical World News,* Aug. 9, 1974, pp. 15–16.

Grobstein, Clifford. "External Human Fertilization." *Scientific American* 240, no. 6 (June 1979): 57–67.

———. *From Chance to Purpose: An Appraisal of External Human Fertilization.* Reading, Mass.: Addison-Wesley, 1981.

Grossman, Edward. "The Obsolescent Mother: A Scenario." *Atlantic Monthly,* May 1971, pp. 39–50.

Hamilton, Michael, ed. *The New Genetics and the Future of Man.* Grand Rapids, Mich.: William B. Eerdmans, 1972.

Harris, Louis. "Brave New World — with Reservations." *Life,* June 13, 1969, p. 52 ff.

Hawthorne, Peter. *The Transplanted Heart.* Johannesburg: Hugh Keartland, 1968.

Henig, Robin Marantz. "Go Forth and Multiply, Ethics Board Tells Scientists." *BioScience* 29, no. 5 (May 1979): 321–23.

———. "In Vitro Fertilization: A Cautious Move Ahead." *BioScience* 28, no. 11 (Nov. 1978): 685–88.

———. "Tempting Fates: If You Could Dictate the Content of Your Kid's Genes, Wouldn't You?" *Discover* 19, no. 5 (May 1998): 58–64.

Hotz, Robert Lee. *Designs on Life: Exploring the New Frontiers of Human Fertility.* New York: Pocket Books, 1991.

Jones, Howard W., Jr. "In the Beginning There Was Bob." *Human Reproduction* 6, no. 1 (1991): 5–7.

———. "Variations on a Theme" (editorial). *Journal of the American Medical Association* 250, no. 16 (Oct. 28, 1983): 2182–83.

Jones, Howard W., Jr., and Charlotte Schrader. "In Vitro Fertilization and Other Assisted Reproduction." *Annals of the New York Academy of Sciences* 541 (Oct. 31, 1988).

Kass, Leon R. "Babies by Means of In Vitro Fertilization: Unethical Experiments on the Unborn?" *New England Journal of Medicine* 285, no. 21 (Nov. 18, 1971): 1174–79.

———. "Making Babies — the New Biology and the 'Old' Morality." *Public Interest,* winter 1972.

———. "'Making Babies' Revisited." *Public Interest,* winter 1979.

Kass, Leon R., and James Q. Wilson. *The Ethics of Human Cloning.* Washington, D.C.: AEI Press, 1998.

Kolata, Gina. *Clone: The Path to Dolly and the Road Ahead.* New York: William Morrow, 1998.

Krauthammer, Charles. "A Secular Argument Against Research Cloning." *New Republic,* Apr. 29, 2002.

Leach, Gerald, *The Biocrats: Ethics and the New Medicine.* Baltimore: Penguin Books, 1970.

Lewis, C. S. *The Abolition of Man.* New York: Macmillan, 1947.

Marsh, Frank H., and Donnie J. Self. "In Vitro Fertilization Moving from Theory to Therapy." *Hastings Center Report* 10 (June 1980): 5–6.

Marx, Jean L. "Embryology: Out of the Womb — Into the Test Tube." *Science* 182 (Nov. 23, 1974): 811–14.

Morell, Virginia. "A Clone of One's Own." *Discover* 19, no. 5 (May 1998): 83–89.

Mydans, Seth. "When Is an Abortion Not an Abortion?" *Atlantic Monthly,* May 1975.

Ozar, David T. "The Case Against Thawing Unused Frozen Embryos." *Hastings Center Report* 15 (Aug. 1985): 7–12.

Pertman, Adam. "Creepy and Inevitable: Cloning Us." *Boston Globe,* Mar. 18, 2001, p. H1

Pincus, G., and E. V. Enzmann. "Can Mammalian Eggs Undergo Normal Development in Vitro?" *Proceedings of the National Academy of Sciences* 20 (1934): 121–22.

Powledge, Tabitha M. "A Report from the Del Zio Trial." *Hastings Center Report* 8, no. 5 (Oct. 1978): 15–17.

Ramsey, Paul. *Fabricated Man.* New Haven: Yale University Press, 1970.

———. "Manufacturing Our Offspring: Weighing the Risks." *Hastings Center Report* 8 (1978): 7–9.

———. "Shall We 'Reproduce'? I: The Medical Ethics of In Vitro Fertilization." *Journal of the American Medical Association* 220, no. 10 (June 5, 1972): 1346–50. "II: Rejoinders and Future Forecast." *Ibid.* 220, no. 11 (June 12, 1972): 1480–85.

Restak, Richard M. "Can There Be New Forms Of Life Before Birth?" *New York Times,* July 16, 1978, p. E8.

Rogers, Michael. *Biohazard.* New York: Alfred A. Knopf, 1977.

———. "The Pandora's Box Congress." *Rolling Stone* 189, no. 36 (June 19, 1975).

Rorvik, David M. "The Embryo Sweepstakes." *New York Times Magazine,* Sept. 15, 1974, p. 17 ff.

———. *In His Image: The Cloning of a Man.* Philadelphia: J. B. Lippincott, 1978.

———. "Making Men and Women Without Men and Women." *Esquire,* Apr. 1969, p. 108 ff.

———. "The Test-Tube Baby Is Coming." *Look,* May 18, 1971, p. 83 ff.

Rosenfeld, Albert. "Challenge to the Miracle of Life." *Life,* June 13, 1969, p. 39 ff.

Rugh, Roberts, and Landrum B. Shettles, with Richard N. Einhorn. *From Conception to Birth: The Drama of Life's Beginnings.* New York: Harper & Row, 1971.

Schulman, Bruce J. *The Seventies: The Great Shift in American Culture, Society, and Politics.* New York: Da Capo Press, 2002.

Schumacher, Gebhard F. B., ed. "In Vitro Fertilization." *Journal of Reproductive Medicine* 11, no. 5 (Nov. 1973): 192–204.

Shettles, Landrum B. "Human Blastocyst Grown in Vitro in Ovulation Cervical Mucus." *Nature* 229 (Jan. 29, 1971): 343.

———. *Ovum Humanum: Growth, Maturation, Nourishment, Fertilization and Early Development.* New York: Hafner, 1960.

Shettles, Landrum B., and David Rorvik. *How to Choose the Sex of Your Baby* (reprint of *Your Baby's Sex: Now You Can Choose*). New York: Doubleday, 1997.

Siebel, Machelle M. "A New Era in Reproductive Technology: In Vitro Fertilization, Gamete Intrafallopian Transfer, and Donated Gametes and Embryos." *New England Journal of Medicine* 318, no. 13 (Mar. 31, 1988): 828–34.

Silver, Lee M. *Remaking Eden: How Genetic Engineering and Cloning Will Transform the American Family.* New York: Avon Books, 1997.

Society for Assisted Reproductive Technology and American Society for Reproductive Medicine. "Assisted Reproductive Technology in the United States: 1998 Results from the SART/ASRM Registry." *Fertility and Sterility* 77, no. 1 (Jan. 2002): 18–31.

Steptoe, P. C., R. G. Edwards, and J. M. Purdy. "Human Blastocysts Grown in Culture." *Nature* 229 (Jan. 8, 1971): 132–33.

Sweeney, William J., III, with Barbara Lang Stern. *Woman's Doctor: A Year in the Life of an Obstetrician-Gynecologist.* New York: William Morrow, 1973.

Taylor, Gordon Rattray. *The Biological Time Bomb.* Cleveland: New American Library, 1968.

"Test Tube Babies." *Reader's Digest* (condensed from *Literary Digest*), Feb. 1937, p. 18.

"Test Tube Babies Pilloried Again." *Nature* 295 (Feb. 11, 1982): 445.

Valone, David A. "The Changing Moral Landscape of Human Reproduction: Two Moments in the History of In Vitro Fertilization." *Mt. Sinai Journal of Medicine* 66, no. 3 (May 1998): 167–72.

Wallis, Claudia. "Quickening Debate over Life on Ice." *Time,* July 2, 1984, p. 68.

Walters, LeRoy. "Ethical Aspects of Surrogate Embryo Transfer" (editorial). *Journal of the American Medical Association* 250, no. 16 (Oct. 28, 1983): 2183–84.

Watson, James D. *The Double Helix.* New York: Atheneum, 1968; reprint, W. W. Norton, 1980.

———. "Moving Toward Clonal Man: Is This What We Want?" *Atlantic Monthly,* May 1971, pp. 50–53.

Watson, James D., and John Tooze, eds. *The DNA Story: A Documentary History of Gene Cloning.* San Francisco: W. H. Freeman, 1981.

Acknowledgments

Reconstructing the stories in this book involved the cooperation of scientists, attorneys, journalists, and others who spent time with me remembering the early days of IVF research: Duane Alexander, Mason Andrews, Jonathan Beard, Jerome Belinson, Lana Shettles Callahan, John Cantu, Doris Del-Zio, John Del-Zio, Michael Dennis, Richard N. Einhorn, Michel Ferin, James Furey, Jane Gardner, Diane Gianelli, William Gibbons, Roger Gosden, Bert Hansen, Lee Hotz, Georgianna Jagiello, Raphael Jewelewicz, Howard Jones, Leon Kass, Lynne Lamberg, Susan Lanzendorf, Markus Leutert, Richard Marr, Luigi Mastroianni, Jacob Mayer, Barbara Mishkin, Jon Palfreman, Tammy Powledge, Kenneth J. Ryan, Christine Soares, Brewer Shettles, Landrum Shettles, Rebecca Skloot, Marcelo Sánchez Sorondo, Sean Tipton, Dominique Toran-Allerand, Betsy Wade, Joyze Zeitz, and Ionnis Zervoudakis. Thanks to all of them, especially to Howard Jones, who did me the huge favor of reading every word of my manuscript to help me make it as scientifically truthful as I could. Thanks too to Martin Rosenberg and the staff of the National Archives in New York, where the Del-Zio trial papers are housed; Judy Engleberg, the archivist at the Joseph P. Kennedy, Jr., Foundation; and Mary Gregg Misch, information specialist at *F-D-C Reports.* And thanks to the Alicia Patterson Foundation, especially Peggy Engel, for giving me a research fellowship to work on a different project at the same time I stumbled upon this one, and generously allowing me the flexibility to follow my interests in a different order than originally intended.

As my first and favorite reader, my husband, Jeff Henig, once again read just about every word of just about every draft. His willingness to discuss the book endlessly was like a benediction; so is his gentle soul. Also a great help were Clare Marantz and Pat McNees for the cheerleading; Anne Derbes and Bob Schwab for the beach house; Judy Kirpich for the lost cause; and David Pollock for the walks in the park.

A few writer friends shared the wisdom only fellow writers can provide. Erik Larson, as brilliant an editor as he is a writer, turned the book inside out and offered suggestions for organization and emphasis that were almost always right. His marginal notes were hilarious, and it was my great good fortune that he also turned out to be a world-class noodge. Thanks too to Deborah Blum, Ellen Ruppel Shell, and Jamie Shreeve. And to everyone at Houghton Mifflin — Anton Mueller, Erica Avery, Megan Wilson, and Peg Anderson — my thanks as well.

Without my peerless agent, David Black, there'd have been no book at all, since I wouldn't have known how resonant this topic could be. I'm grateful to have found David and to have benefited from his smarts, his perseverance, his spirit, and his overarching decency.

And finally, my girls. Writing about the desolation of couples unable to have children reminded me how precious are my own two daughters, Jessica and Samantha. As Jeff and I have watched them grow into remarkable young women, we've learned that the searing love we feel for them is what baby making, whether high-tech or low, is all about.

INDEX

"Abolition of Man, The" (Lewis), 14
Abortion: and cloning, 249–50; Del-Zio's, 115; legalization of, 6, 220; and Norfolk IVF clinic, 226–27; opponents of, 11, 59–60, 79–81, 129, 206–14, 223, 253. *See also* Embryos; Fetuses
Abram, Morris, 211
ACT (Advanced Cell Technology), 246
Adams, Peter, 87
Adoption, 259
Advanced Cell Technology (ACT), 246
"After Making Love We Hear Footsteps" (Kinnell), 256
AID. *See* Artificial insemination
Albert and Mary Lasker Awards, 45
Alicia Patterson Foundation, fellowship for Rorvik, 147
Allen, Woody, 143
Allerand, Dominique Toran-, 23–25, 97, 102–3, 196
Amazon Odyssey (Atkinson), 6
American Civil Liberties Union, 129
American College of Obstetrics and Gynecology, 213
American Medical News, 212
American Society for Experimental Pathology, 128
American Society for Pediatric Research, 87
Amherst College, 140
Amniocentesis, 11, 226–27
Andrews, Lori, 273n
Andrews, Mason, 92, 139, 140, 175, 204–6, 214, 223–25
Andromeda Strain, The (Crichton), 146
Animal Biology, 52
Animals: IVF research on, 2, 29–30, 33–34, 40, 42, 62–63, 96, 130–31, 137, 174, 273n, 279nn; need for more research on, 10, 58, 61–63, 121, 187, 246; as poor models for human reproduction, 62–63, 74, 137, 187; transplanting of hearts from, into humans, 10
Antibiotics, 83, 84, 111–12, 286n
Apollo 13 space flight, 157
ART (assisted reproductive technology). *See* Reproduction (assisted)
Artificial insemination (AID): Doris Del-Zio's attempts at, 1, 115; history of, 26–29, 174; ignorance about, 67–68; number of U.S. births associated with, 29, 68; slippery slope arguments about, 10, 50, 70; "test tube babies" as born through, 67–68
Artificial wombs, 32, 211
Asilomar conference (California), 121–28, 258, 287nn
As Man Becomes Machine (Rorvik), 147
Assessing Biomedical Technologies (National Academy of Sciences), 174–75
Assisted hatching, 234
Assisted reproduction. *See* Reproduction (assisted)
Assisted suicide, 11

Association of American Medical Colleges, 129
Association of Professors of Gynecology and Obstetrics, 184
Atkinson, Ti-Grace, 6
Atlantic Monthly, 67, 68, 74–76
Atomic Age, 39, 157, 211, 290n
Atomic bomb, 74, 86, 257, 265
Australia: abortion opponents from, 214; embryo freezing in, 241; IVF research in, 217–18, 238, 239; test tube babies born in, 203, 210, 212, 218, 224, 228

Babies, desire for, 104–5, 108–9, 115, 259–60
Baltimore, David, 126
Barnard, Christiaan, 37–38, 47–48, 277n
"Battle of the Sexes" tennis match, 105–6
Bavister, Barry, 78
Beckwith-Wiedemann syndrome, 240, 241, 300n
Belgium, 24, 297n
Belmont report, 255
Berg, Paul, 118, 119, 124–25
Bernstein, Leonard, 64–65
Bevilacqua, Anthony, 10
Bevis, Douglas C. A., 112–14, 157, 246
Biggers, John, 274n, 282n, 297n
Binkert, Alvin J., 272n
Bioethics: commissions on, 11–12, 80–81, 90–91, 111–12, 128–32, 211, 253–57; and Eunice Kennedy Shriver, 45–47, 66, 88, 91; and fetal research, 11–12, 80–81, 88–90, 127–32; and genetic engineering, 116–27, 255, 257; growth of, 5, 45–47, 64–77, 82–84, 88–90, 99–100, 127–32, 135–38, 252–57; guiding principles of, 255; issues discussed in, 64–77, 99–100, 137–38. *See also* Consent forms; Ethics Advisory Board; Human experimentation; Knowledge; Regulation of medical research; Science: concerns about dark side of; Slippery slope arguments; Technology assessments
Bionic Man (on television show), 109–10
Birth control, 6, 52, 220
Birth defects: Beckwith-Wiedemann syndrome, 240, 241, 300n; from cloning, 252; IVF and early fears of, 10, 13, 73, 95, 97, 98, 120–21, 128, 130–32, 135, 136–37, 150, 168, 169, 173, 201, 210; IVF and recent evidence of, 238–43, 299–300nn; low birth weight, 239, 300n; bladder exstrophy, 241, 300n; as miscarriage cause, 98, 150
"Birthmark, The" (Hawthorne), 256
Birth weight, low, 239, 300n
Bishop, Jerry, 146
Bladder exstrophy, 241, 300n
Blandau, Richard, 178–79, 187
Blastocysts: defined, 2, 248, 250; Edwards and Steptoe's growing of, 72, 78–79; implantation of, into uterine wall, 2–3, 36, 62, 113, 181–82, 279n; as potential human being, 10; Shettles's photos of, 20, 35, 50–53, 55, 60, 67, 178–79, 278n. *See also* Embryos; Morulas; Zygotes
Blood transfusions, 174
Boalt Hall School of Law (Berkeley, California), 139
Boisselier, Brigitte, 245–47
Bok, Sissela, 135
Bone marrow transplants, 38
Book of Popular Science, The, 52
Bourn Hall fertility clinic (Great Britain), 222
Bower, John, 155–56, 158–59, 164–65, 168, 178–80, 183, 186
Boyle, Lord, 114
Boys from Brazil, The (Levin), 250
Brackett, Benjamin, 173–74
Brain, fetal, experiments on, 88
Brain death, 11, 48–49
Brain disease, and fetal tissue, 254

Brave New Baby (Rorvik), 147
Brave New World (Huxley), 72, 143, 251
Brecht, Bertolt, 193
Breckon, William, 143
British Medical Association, 112–14
Brown, Casey, 262
Brown, John, 133–34, 141, 170–71, 230, 262, 288n
Brown, Lesley, 133–34, 140–41, 230, 262; pregnancies of, 149–52, 157, 158, 170–72, 230
Brown, Louise: birth of, 170–72, 176, 177, 201, 242; book about, 262, 288n; as first test-tube baby, 5, 202, 238; health of, 187, 224, 229; sister of, 230
Brown, Natalie, 230, 262
Brown, Sharon, 133
Buckley, James L., 91
Bush, George W., 255–57
Bush, Vannevar, 173
Byron, George Gordon, Lord, 8, 9

Cajal, Santiago Ramon y, 102–3
Califano, Joseph, 134–35, 188, 223
Callahan, Daniel, 66, 279n
Callahan, Lana Shettles, 263, 277n, 278n
Cambridge University (England), 31, 32, 43, 69, 74, 78
Canada, 236
Capron, Andrew, 123–24
Carr, Elizabeth Jordan, 226, 238, 264, 302n
Carr, Judy, 220–26
Carr, Roger, 220, 221, 224, 225
Carson, Johnny, 147, 189–90
Carter, Jimmy, 134, 207, 223
Case Western Reserve University (Cleveland), 87
Catholic church: "ensoulment" theory of, 81; on "fetal decapitation," 88; on fetal research, 129; on IVF, 10, 31, 52–53, 134–35, 231
Catholic Youth Encyclopedia, 52
CDC (Centers for Disease Control), 233, 235, 237, 299n
Center for Bioethics (Georgetown University). *See* Kennedy Institute for Bioethics
Center for Evaluation of Clinical Procedures, 89
Center for Law and Social Policy, 89
Centers for Disease Control (CDC), 233, 235, 237, 299n
Certificates of need, 207–10
Chancellor, John, 68, 280n
Chang, Min-Cheuh, 33–34, 130–31, 279n
Charo, Alta, 246
Child Is Born, A (Nilsson), 53; number of, from IVF, 226, 228
Children's Defense Fund, 89
Chimeras: fears about creating, 138, 258; IVF techniques applied to, 11; through genetic engineering, 118, 132, 215–16
Choice (in bioethics), 100
Clinton, Bill, 254, 255
Clonaid, 245
Cloning: attempts to outlaw, 12, 74, 248–49, 252, 256–57; concerns about, 237–38, 248–52, 255, 258–59, 301n; human, claims about, 138, 144–49, 245–47, 262–63, 266; *In His Image* (Rorvik), 144–47, 149, 289–90nn; and IVF, 10–14, 142–43, 148, 258–59; Kass on, 10, 70; possible benefits of, 258–59; process involved in, 13, 142–43; of sheep, 248, 252, 255, 301n; "therapeutic" vs. "reproductive," 247–50, 256–57, 266; Watson on, 74–76. *See also* Stem cell research
CNN, 246
Coalition for Life, 129
Cohen, Jean, 222–23
Cohen, Stanley, 125–26
Cohn, Victor, 288n
Coleman, Marshall, 209–10
Columbia Presbyterian Medical Center.

See Columbia University's College of Physicians and Surgeons; Presbyterian Hospital (New York City)
Columbia University's College of Physicians and Surgeons, 89; Del-Zios' suing of, 158; fertility clinic at, 230–31; letters of assurance at, 59; on Shettles's sex selection methods, 60. *See also* Presbyterian Hospital (New York City)
Commission ethics, 252–53
Committee on the Life Sciences and Social Policy, 70
Conan Doyle, Arthur, 56
"Conception in a Watch Glass" (unsigned article), 29–30
Congress: and cloning, 74, 76, 147–48; and fetal research, 116, 128–32, 188, 254–55; and scientific oversight, 124–26, 193, 253–54. *See also* Government funding; Politicians
Consent forms: for Del-Zios' IVF, 3–4; as regulatory mechanisms, 58
Consumer Age, 39, 157
Contraception. *See* Birth control
Cornea transplants, 38
Cornell Medical School, 3, 231. *See also* New York Hospital
Crichton, Michael, 146
Crick, Francis, 32, 74
Crow, James, 279n
Cryopreservation (of embryos), 63, 136, 241–42, 300nn
Culdoscopy, 45
Cummings, E. E., 245
Curettage, 44–45

Daily Mail (London), 114, 141, 170–71
Darvall, Denise, 47
Dean, Charles, Jr., 206–10, 213–14
Death, 11, 48–49
Declaration of Helsinki, 58–59
Decompression Babies (Rorvik), 147
De Kruif, Paul, 26, 155
Del-Zio, Debbie, 2
Del-Zio, Denise, 2
Del-Zio, Doris, 133, 212; after IVF attempt, 105, 107–12, 114–16, 166; after trial, 261; artificial insemination attempts of, 1, 115; first husband's abuse of, 115, 165–66; IVF attempt by, 1–5, 19–20, 23–25, 95, 96, 98, 100–102, 169, 177, 181, 183, 190–92, 229–30; lawsuit of, 115–16, 155–69, 177–96, 197, 201, 230, 261, 269n; as trial witness, 159–67
Del-Zio, John: after IVF attempt, 110–11; after trial, 261; artificial insemination attempts of, 1, 115; lawsuit of, 115–16, 155–69, 177–96, 197, 201, 230, 261, 269n; role of, in wife's IVF, 1–5, 19–20, 23, 56
Del-Zio, Tammy, 2, 101, 109, 110, 115, 164, 261
Del-Zio v. Vande Wiele et al., 115–16, 155–69, 177–96, 201, 230, 261, 269n
Dennis, Michael, 116, 158–59, 161, 165, 178, 180, 181, 183, 185–87, 192
Department of Health, Education, and Welfare (HEW): clinical research guidelines from, 59; and Ethics Advisory Board, 130–32, 134–35, 223; and fetal research, 91, 130–32, 254; on Tuskegee Syphilis Study, 83–84. *See also* Ethics Advisory Board
Designer babies, 215
Dictionary of Biology, 52
Divorce: and artificial insemination, 28; and infertility, 26; no-fault, 6
DNA: discoverers of structure of, 32, 74; recombinant, 116–27. *See also* Genetic engineering
Dolly (cloned sheep), 248, 252, 255, 301n
Doornbos v. Doornbos, 28, 29
Dostoyevsky, Fyodor, 233
Double Helix, The (Watson), 74
Down's syndrome, 65
Drugs. *See* Antibiotics; Birth control; Fertility drugs

Dubos, René, 150
Dunaway, Faye, 24
DuVal, Merlin, 83–84

Eastern Virginia Medical School (Norfolk): Andrews and, 92, 139, 175, 204, 205, 206, 223; creation of, 92, 139; as first IVF clinic in U.S., 91–92, 203–14, 217–23, 230, 264, 270n; Joneses' employment at, 139–40, 175–76
Eccles, John, 68
Ectogenesis, 70
Ectopic pregnancies, 220, 238
Edelin, Kenneth, 220, 297n
Edwards, Caroline, 32–33
Edwards, Jennifer, 32–33
Edwards, Robert, 231, 242; background of, 32–33, 68; called "baby killer," 76, 77, 88; ethical concerns of, 97–98, 121; family of, 32–33, 40; fertility clinic of, 222, 228; and first test tube baby's birth, 170–72; IVF research of, 32–35, 39–44, 49–50, 61–63, 113, 137, 150–51, 160, 181, 188–89, 202, 210, 229; and Joneses, 40, 41–44, 222, 276nn, 295n; at Kennedy Foundation conference, 69–71, 74, 76–77; later career of, 264; motives of, 32, 38, 40, 49–50; as "mouse geneticist," 40, 62; and Steptoe, 49–50, 69, 71, 78, 150, 170–72; Vande Wiele's view of, 96
Edwards, Ruth, 32
Eggs (ova): customized cloning of, 258–59; Doris Del-Zio's, for IVF, 4, 20; fertilization of, by more than one sperm, 34, 82, 202; fertilization of, outside human body, 29–31, 43; Jagiello's study of, 24, 190–92; retrieval of, for IVF, 2, 33, 39, 41, 43, 45, 49–50, 71, 77, 276n; Shettles's photographs of, 20, 35, 51–53, 55, 60, 67, 178–79, 278n; Shettles's study of, 35–39; as tissue, 35; waste of, in nature, 82. *See also* Blastocysts; Fertilization; Gametes; Implantation; Meiosis; Morulas; Oocytes; Sperm; Zygotes
Ehrlich, Paul, 108
Einhorn, Richard N., 53
Einstein, Albert, 78
Eiseley, Loren, 177
Embryos: defined, 250; destruction of, 35, 76, 81–82, 136, 138, 161, 163–64, 205, 206; frozen, 63, 136, 241–42, 300nn; government policies on, 11, 80–81, 90, 188, 231, 253–55, 257; informed consent and, 71, 91, 121, 129; and IVF multiple births, 235–37, 239; IVF techniques for transferring, 11, 36–37, 71, 113; laboratory creation of, 10–12, 31–32, 55, 205, 247–48, 254, 257; moral status of, 10, 81, 88, 129, 136; regulations for work with, 89–91, 121, 205, 254, 257; Soviet claims about, 32. *See also* Blastocysts; Cloning; Eggs; Fetuses; Implantation; In vitro fertilization; Morulas; Sperm; Stem cell research; Zygotes
England. *See* Great Britain
Environmental movement, 6–7, 105, 108
Escoffier-Lambiotte, Claudine, 280n
Esquire, 56, 147
Estes operation, 213
Ethics. *See* Bioethics; Situational ethics
Ethics Advisory Board (HEW): Califano and, 134–35, 223; establishment of, 130–38; hearings held by, 135–38; on IVF, 205, 206, 211, 223, 253; Soupart and, 130–31, 136–37, 205, 223–24
Eugenics Society (Great Britain), 28

Fallopian tubes: blocked, 1, 36, 72–73, 109, 115, 160, 213; diseased, 285n; ectopic pregnancy's effects on, 220; IVF's bypassing of, 40, 49, 212; of mice, 62. *See also* Eggs; Infertility; Sperm

Fallopio, Gabriele, 62
Family Circle, 54
Federal funding. *See* Government funding
Female Eunuch, The (Greer), 107
Feminism, 6, 105–7, 128–29
Ferin, Michel, 232, 271nn, 298n
Fertility clinics: first U.S., 91–92, 203–14, 217–23, 230, 264, 270n; in Great Britain, 222, 228; and multiple births, 237; number of, 230, 233; pregnancy rates of, 236, 299n; private funding of, 12, 123, 176, 203–6. *See also* Australia; Infertility; In vitro fertilization: entrepreneurial; Pregnancies
Fertility drugs: Carr's taking of, 220–21; Del-Zio's taking of, 3, 4; with IVF, 49, 71, 78, 113, 219–21, 235; risks associated with, 235, 239, 243
Fertilization (artificial): difficulties of, in IVF, 50, 84–86; Edwards's and Jones's achievement of, 43; Edwards and Steptoe's witnessing of, 50–51; IVF techniques for, 11, 49, 71–72, 113; Shettles's claims of, 35–39, 43, 275n; signs of, 43. *See also* Blastocysts; Eggs; Embryos; Implantation; Morulas; Reproduction (assisted); Sperm; Zygotes
Fetuses: "decapitated," 87–89, 131; defined, 250; government policies on, 11–12, 80–82, 91, 111–12, 116, 128–32, 211, 231, 253–54; moral status of, 81, 129; Soviet claims about, 32, 273n. *See also* Embryos
Feversham Report (Great Britain), 28
Fire (in Prometheus myth), 8, 9, 15, 47
Fletcher, John, 99–100, 128, 283n
Fort Detrick (Maryland), 127
France, 222–23, 238
Francoeur, Robert T., 280n
Frankenstein (Shelley), 1, 8–9, 47, 100, 104–5
From Conception to Birth (Rugh and Shettles), 53
Furey, James, 163–64, 166–67, 181–83, 274n, 275nn, 291nn
"Future of Man, The" conference (Paris), 120–21

Gairdner Foundation, 302n
Gaither, James, 135
Galileo, 55, 71, 77
Gallup polls, 190
Gamete intrafallopian transfer (GIFT), 234, 238, 272n
Gametes: definition of, 21; freezing of Del-Zios', 100; Pincus's work with, 29; testing for risks associated with lab uses of, 130–31, 136–37. *See also* Eggs; GIFT; Sperm
Garbo, Greta, 24
Garcia, Jairo, 222, 223
Gattaca (movie), 256
Gaull, Jerald, 80
Gaylin, Willard, 19, 66
Gene splicing. *See* Genetic engineering
Genetic diagnosis (preimplantation), 11, 231, 257
Genetic engineering (gene splicing): benefits of, 257–58; congressional interest in, 74; ethical issues in, 122–24, 126–27, 255, 257; fears about, 211, 258; IVF different from, 117, 132; and IVF, 11; of mice, 215–16, 257–58; scientists' moratorium on, 116–27. *See also* Cloning
Georgetown University, 66, 80, 83, 89, 128
Gibbons, Bill, 296n
Gifford Memorial Hospital (Randolph, Vermont), 262–63, 289n
GIFT (gamete intrafallopian transfer), 234, 238, 272n
Glamour, 264
Gonorrhea, 26–27
Good Health, 52
Good Housekeeping, 261

Gordon, Jon, 215–16, 257–58
Goucher College (Maryland), 140
Government funding: advantages and disadvantages of, 61, 79; and fetal research, 88, 253; IVF's lack of, 11–12, 61–63, 79–80, 89–90, 130, 134–35, 205–6, 211, 223, 230; loss of, for violations of human experimentation procedures, 59–60, 95, 97, 98, 101–2, 185, 188; as regulatory mechanism, 12, 59, 79, 223. *See also* Congress; Market motives; National Institutes of Health; Politicians
Great Britain: claims about IVF successes in, 112–14; first test tube baby born in, 5, 133–41, 158–59, 202–3, 210, 262; government funding for research in, 61, 63; IVF quintuplets in, 236; IVF studies in, 238; opinions about reproductive technologies in, 28, 50, 188. *See also* Cambridge University; Edwards, Robert; Oldham; Steptoe, Patrick
Greek myths, 7–9, 14–15, 47, 71
Greer, Germaine, 68, 107
Grobstein, Clifford, 273n
Guy's Hospital (London), 191

Hall, Kristin, 106, 285n
Halpern, Philip, 188
Hamburg, David, 135
Ham's F-10 culture medium, 78, 85
Hamsters, 63, 78, 279n
Handler, Philip, 68
Hanford nuclear weapons plant, 290n
Hard, Addison Davis, 272n
Harris polls, 50, 68, 201
Harvard Standards, 48–49
Harvard University, 29–31, 33, 48–49
Hastings Center (Institute of Society, Ethics, and the Life Sciences), 66, 89, 137, 279n
Hatch, Orrin, 209–11
Hawthorne, Nathaniel, 256
HCG (human chorionic gonadotropin), 78
Heart: animal-to-human transplants, 10; defects of, 238, 243; transplants of, 10, 11, 37–38, 116, 277n
Hellegers, André, 80
Helsinki Declaration, 58–59
HERP (Human Embryo Research Panel), 80–81, 254
HEW. *See* Department of Health, Education, and Welfare
Hoffman, Eva, 251
Hoppe, Peter, 148
Host-vector pairs (in molecular biology), 122–23, 126–27
Humanae Vitae (Paul VI), 52–53
Human Embryo Research Panel (HERP), 80–81, 254
Human experimentation: ethical concerns about, 58–59, 74, 82–84, 90–91, 97, 128–30, 168–69, 283n; loss of funding for procedural violations of, 59–60, 95, 97, 98, 101–2, 185, 188. *See also* Animals; Bioethics; Cloning; Embryos; Fetuses; In vitro fertilization; Regulation of medical research; Reproduction (assisted)
Human Fetal Tissue Transplantation Research Panel (NIH panel), 254
Human Genetics and Its Foundations, 52
Huxley, Aldous, 72, 143, 250
Hysterectomy, 36, 55, 181
Hysterotomy, 87–88

ICSI (intracytoplasmic sperm injection), 240–41, 243, 300n
Ignorance. *See* Knowledge
Illegitimacy (and artificial insemination), 28, 29
Illmensee, Karl, 148
Implantation (of fertilized eggs into uterus), 2–3, 36, 49, 62, 72, 113, 181–82,

279n; genetic diagnosis before, 11, 231, 257; Shettles's claims about, 35–39, 181–82, 274n, 275n. *See also* Transplantation
Incubators: and Del-Zios' IVF attempt, 23, 191, 192; in IVF, 2, 71, 78
India, 203, 210, 294–95n
Infanticide, 76
Infertility: causes of, 136, 213, 220, 285n; Edwards's concerns about, 32–33, 40, 49–50; IVF as solution for, 30; IVF not "treatment" for, 72–74, 213; and longing for children, 104–5, 108–12, 115, 166, 259–60; number of couples experiencing, 108–9, 214–15, 285n; number of physicians specializing in, 107; secondary, 115; seen as woman's fault, 26; Steptoe's concerns about, 44–45. *See also* Fallopian tubes; Fertility clinics; In vitro fertilization; Reproduction (assisted)
Information Age, 39, 157
Informed consent: in 1950s and 1960s, 35, 36; for postabortion fetal research, 111–12; and research on human embryos, 71, 91, 121, 129
In His Image: The Cloning of a Man (Rorvik), 144–47, 149, 289–90nn
Institute of Experimental Biology (Moscow), 32
Institute of Society, Ethics, and the Life Sciences. *See* Hastings Center
Insurance (medical), 6, 233
Intracytoplasmic sperm injection (ICSI), 240–41, 243, 300n
In vitro fertilization (IVF): animal research on, 2, 29–30, 33–34, 40, 42, 62–63, 96, 130–31, 137, 174, 273n, 279nn; benefits of knowledge about, 252, 265; Bevis's claims of, 112–14, 157, 246; and birth defects, 10, 13, 73, 95, 97, 98, 120–21, 128, 130–32, 135, 136–37, 150, 168, 169, 173, 201, 210, 235–43, 299–300nn; changes in views about, 4–7, 12, 13, 15, 173–77, 201, 229–34, 237–38; and cloning, 10–14, 142–43, 148, 258–59; Del-Zio's attempt at, 2–5, 19–20, 23, 169, 177, 229–30; entrepreneurial, 12, 176, 203–6, 214–15, 219–23, 230–32, 237, 241–42; ethical concerns about, 6, 9–12, 50, 70, 73, 86, 117, 127–32, 264–66; first child born of, 133–34, 140–41, 150–52, 170–72, 177; first U.S. clinic for, 91–92, 203–14, 217–23, 230, 264, 270n; history of development of, 29–45, 49–51, 61–63, 69–70, 78–79, 84–86, 133–34, 140–41, 150–52, 170–77, 202–3, 212–14, 217–23, 229, 233, 264, 265–67; lack of government funding for, 11–12, 61–63, 79–80, 89–90, 130, 134–35, 205–6, 211, 223, 230; multiple births associated with, 235–37, 239; number of annual procedures of, in U.S., 233; process involved in, 2–3, 36, 113, 218–21, 235; pros and cons of, for couples, 242–44; public costs associated with, 243–44; regulatory mechanisms for, 5, 11–12, 135, 223; as routine procedure, 6, 228, 233–35, 238–39, 242; Shettles and, 35–39, 181–82; success rate of current, 233, 235; technology of, 11, 231–32, 264; variations on, 234, 240–41. *See also* Bioethics; Blastocysts; Eggs; Embryos; Fertility clinics; Fertility drugs; Fertilization (artificial); Implantation; Publicity; Regulation of medical research; Reproduction (assisted); Science; Slippery slope arguments; Sperm; Zygotes
In vivo fertilization, 29, 219
Island of Dr. Moreau (Wells), 258
Israel, 238
Italy, 31–32
IVF. *See* In vitro fertilization
IVF Mafia, 204

"I Was Cheated of My Test Tube Baby" (Del-Zio), 261

Jagiello, Georgianna, 24, 25, 190–92, 195–96, 215, 230, 232
Jaroff, Leon, 146
Jefferson Medical College, 26
Jewelewicz, Raphael, 298n
Johannesburg Sunday Express, 48
Johns Hopkins University, 89; documentary produced at, 65; IVF studies at, 240, 241; Joneses' affiliation with, 22, 40–43, 138–40
Jonathan Livingston Seagull (Bach), 67
Jones, Georgeanna, 231, 276n; background of, 42, 139–40, 297n; at Eastern Virginia Medical School, 175–76, 203–4, 206, 213, 214, 217–23, 225–26, 264, 270n; and Edwards, 40, 41–44, 222, 276nn, 295n; later career of, 263–64; on motherhood, 42, 227–28
Jones, Howard W., Jr., 231, 276n, 279n; background of, 42; at Eastern Virginia Medical School, 175–76, 203, 204, 206, 213, 214, 217–26, 264, 270n; and Edwards, 40, 41–44, 222, 276nn; and IVF's multiple births, 237; at Kennedy Foundation conference, 69–71, 77; later career of, 263–64; on Pincus, 272n; on Shettles, 22
Jones Institute for Reproductive Medicine (Norfolk, Virginia), 264, 270n. *See also* Eastern Virginia Medical School
Jonsen, Albert, 128
Joseph and Rose Kennedy Institute for the Study of Human Reproduction and Bioethics (Georgetown University), 66, 80, 83, 89, 128
Joseph P. Kennedy, Jr., Foundation conference on bioethics, 65–78, 88
Journal of Fertility and Sterility, 179
Journal of Reproductive Medicine, 22, 178–79

Kahn, Herman, 118
Kass, Leon: as bioethicist, 128, 174–75, 280n, 292–93n; as cloning opponent, 10, 255; and fetal research, 89; as IVF opponent, 10, 70–71, 73, 74, 76, 77, 137–38, 229, 255; and stem cell research, 255–57
Kenley, James B., 209
Kennedy, Edward M., 67, 90–91, 125–26, 128, 211
Kennedy, John F., 64, 65
Kennedy, Rosemary, 66
Kennedy Center for the Performing Arts (Washington, D.C.), 64–65, 68
Kennedy Foundation conference on bioethics, 65–78, 88
Kennedy Institute for Bioethics (Georgetown University), 66, 80, 83, 89, 128
Kershaw's Cottage Hospital (Royton, England), 85, 86
Kidney transplants, 38, 48
King, Billie Jean, 105–7
Kinnell, Galway, 256
Klemesrud, Judy, 230, 231
Knowledge: drive for, 5, 14, 15, 22, 30–31, 76–77, 99–100, 123, 134, 136, 173, 182, 206, 218, 265–67; fear of, 5, 14, 15, 30–31, 70–71, 86, 99, 136, 201, 206, 283n; gained from IVF research, 252, 265. *See also* Bioethics; Science; Scientists; Slippery slope arguments
Knowles, John, 68
Krauthammer, Charles, 249–50
Krol, John, 88

Landmark Communications, 227
Lanza, Robert, 246
Laparotomy, 45, 97
Laparoscopy, 44–45, 49–50, 71, 77

Las Vegas (Nevada), 263
Lawless, Hugh, 166, 194–96
Leach, Gerald, 278n
Lederberg, Joshua, 68, 124–25, 280nn
LeJeune, Jerome, 280n
Letter of assurance: as regulatory mechanism, 59, 95, 278n; and Shettles's IVF experiments, 185, 188
Levin, Ira, 250
Lewis, C. S., 13–14, 75, 142
Life. creation of; as essence of IVF, 9–12; creation of, as Promethean trick, 7–9. *See also* Bioethics; Reproduction (assisted); Reproduction (sexual)
Literary Digest, 27–28
Liver transplants, 38
Look magazine, 54–56, 60, 63, 67, 147
Loren, Sophia, 24
Lowe, Charles, 80
Lynch, Linda, 222

Mailer, Norman, 20
Majors, Lee, 109
Mammals, 62. *See also* Animals; *and see specific mammals*
Market motives: and frozen embryos, 241–42; and human cloning, 266; and IVF, 91–92, 108–9, 203–5, 214–15; and multiple births, 235–37; and organ transplants, 37–38. *See also* In vitro fertilization: entrepreneurial
Markle scholarship, 21
Marshall, Edith, 171
Maryland Action for Human Life, 129
Mass (Bernstein), 64–65
Massachusetts: abortion in, 220, 297n; fetal research in, 111–12; IVF in, 220
Mastroianni, Luigi, 187, 232
Mays, Stephen and Amanda, 218, 296n
McCarthy, Charles, 188
McCormick, Richard, 135
McGee, Frank, 68
McHugh, James, 129
McKusick, Victor, 40, 42
Medical College of Virginia (Richmond), 92
Medical Research Council of Great Britain, 61, 63
Medical World, 27, 272n
Medical World News, 113
Meier, Renee, 88
Meiosis, 41, 43, 82
Melbourne (Australia), 203, 210, 212, 214, 224, 228, 231, 236
Mencken, H. L., 140
Menkin, Miriam, 30, 31
Mental retardation, 45–46, 65, 66, 80, 91
Mice: cloning of, 148, 215–16; and IVF, 2, 31, 33, 40, 62–63, 74, 273n; "transgenic," 215–16, 257–58
Milton, John, 133
Ministry of Health (Great Britain), 171, 172
Mintz, Beatrice, 146
Miscarriages: Doris Del-Zio's, 1, 115, 165–66; as nature's way of dealing with birth defects, 98, 150; and IVF, 202, 222–23, 238, 295–96n
Mitchell, Glenn, 207
Mitgang, Herbert, 147
Molecular biology, 118, 122–23, 126–27, 132. *See also* Genetic engineering
Monash University (Melbourne, Australia), 217–18, 224, 228
Monkeys, 42, 61, 74, 187. *See also* Primates
Monod, Jacques, 68
Moore, J. George, 22
Morgan, Marabel, 107–8
Morulas: defined, 2, 250; Edwards's growth of, 79; Shettles's claims about growing, 35, 179. *See also* Blastocysts; Embryos; Zygotes

"Morula Stage of Human Ovum Developed in Vitro, The" (Shettles), 179
Motherhood: Doris Del-Zio's desire for, 105, 107–8; Georgeanna Jones on, 42, 227–28; longing for biological, 104–5, 108–12, 115, 166, 259–60; as obsolete, 74; as optional, 6, 107–8; surrogate, 50, 75, 135, 266 (*see also* In vitro fertilization; Pregnancies)
"Moving Toward Clonal Man" (Watson), 74–76
MRC (Medical Research Council of Great Britain), 61, 63
Mudd, Roger, 68
Muir, Brett, 236
Muir, Christopher, 236
Muir, Mrs., 236
Multiple births (from IVF), 218, 235–37, 239
Multiplicity (movie), 250

National Abortion Rights Action League, 128–29
National Academy of Sciences, 70, 118, 135, 137, 174–75, 215–16
National Bioethics Advisory Commission, 254–55
National Cancer Institute, 61
National Commission for the Protection of Human Subjects of Biomedical and Behavioral Research, 91, 111–12, 128–30, 211, 253
National Environmental Protection Act, 7
National Institute for Child Health and Human Development, 80
National Institutes of Health (NIH): ethical standards for research under, 45, 88–89, 97, 98, 101–2, 126–27; on IVF research, 80–81; Kass at, 70; legislation for, 254; and human experimentation, 59–60, 95, 97, 98, 101–2, 185, 188; protest at, 88; and "war on cancer," 61. *See also* Government funding; Regulation of medical research
National Youth Pro-Life Coalition, 129
Nature, 71, 216
Nazis, 25, 80, 186, 293n; and Vande Wiele, 25, 186, 271n
New England Journal of Medicine, 29–30, 98, 216, 239, 286n
Newsweek, 146
New York City Ballet, 24
New York Daily News, 164, 166, 174
New York Hospital: as branch of Cornell Medical School, 231; Doris Del-Zio in, 1, 3, 100–102, 105, 106, 160, 229; human experimentation committee at, 168
New York magazine, 54, 147
New York State Institute for Basic Research in Mental Retardation, 80
New York Times, 82–83, 146–47, 226, 230, 231, 264
New York Times Book Review, 146
New York Times Magazine, 147
New York World's Fair, 39, 157
NIH. *See* National Institutes of Health
NIH Revitalization Act of 1993, 254
Nilsson, Lennart, 53
Nixon, Richard, 61, 64, 65, 81, 91
Nobel Prize, 45, 68, 74, 102
Noonan, John, 280n
Norfolk (Virginia), 92. *See also* Eastern Virginia Medical School
Northwestern Memorial Hospital (Chicago), 106–7, 285n
Nova, 86
Nuclear power. *See* Atomic Age

Ob/Gyn News, 87, 187
Oedipus, 71
Oldham (Lancashire, England), 44–45, 69, 78, 85, 86, 170–72, 231

O'Leary, Stephen, 161–63, 184, 185
Omni magazine, 147
Oocytes, 33–36, 39, 41. *See also* Eggs
Orwell, George, 260
Osler, William, 95
L'Osservatore Romano, 31
Our Miracle Called Louise (Brown), 262, 288n
Ova. *See* Eggs
Ovarian hyperstimulation syndrome, 235
Overpopulation, 6–7, 108
Oversight. *See* Regulation of medical research
Oviducts. *See* Fallopian tubes
Ovulation, 62–63. *See also* Eggs; Fallopian tubes; Fertility drugs
Ovum Humanum (Shettles), 20, 51–52

Paese Sera, 31
Pancoast, William, 26–27
Pandora myth, 14–15
Parkinson's disease, 254
Parshley, Mary, 20–21, 23, 25, 102, 103, 191, 192
Parthenogenesis, 43
Pastore, Pierfranco, 10
Paul VI (pope), 52–53
Pedersen, Roger A., 269n
Penicillin, 83, 84
Pergonal (fertility drug), 78, 220–21
Perone, Nicola, 272n
Petrucci, Daniele, 31–32, 35, 246
Phillip, Elliott, 290n
Photographs: Edwards's, 43; Shettles's, 20, 35, 51–53, 55, 60, 67, 178–79, 278n; Steptoe's, 49
Pincus, Gregory, 29–30, 272n
Pius XII (pope), 31
Plasmids, 122, 125–26
Podhoretz, Norman, 68
Politicians: as bioethics watchdogs, 90–91; and cloning, 12, 252–53; and embryo research, 11–12, 81. *See also* Congress
Polson, Carla, 224
Polson, John, 224
Polycystic ovarian syndrome, 33, 41, 49
Population Bomb, The (Ehrlich), 108
Pratt, Lawrence D., 210, 295n
Predictions (scientific), 75, 257–59, 265–67
Pregnancies: and antibiotics, 111–12, 286n; ectopic, 220, 238; Judy Carr's IVF, 222; Lesley Brown's IVF, 149–52, 157, 158, 170–72; Linda Reed's IVF, 212; miscarriage rate with IVF, 238; percentage of difficulties in, 151; rates of, with current IVF, 233, 235–37, 299n. *See also* Abortion; Miscarriages; Motherhood
Presbyterian Hospital (New York City): Del-Zios' suing of, 116, 155–69, 177–92; Eastern Virginia Medical School vs., 203–5; Shettles's relations with, 3, 4, 19–24, 56–57, 95, 97–98, 103, 116, 180, 203–5. *See also* Columbia University's College of Physicians and Surgeons
President's Commission for the Study of Ethical Problems in Medicine and Biomedical and Behavioral Research, 211. *See also* National Commission for the Protection of Human Subjects of Biomedical and Behavioral Research
President's Council on Bioethics, 255–57
Primates, 63, 121, 137. *See also* Monkeys
Proceedings of the National Academy of Sciences, 215–16
Profits. *See* Market motives
Prometheus myth, 7–9, 14, 15, 47, 100
"Prometheus" (Byron), 8, 9
Pronuclei, 43

Public Health Service, 83
Publicity: about Bevis's IVF claims, 113, 157; about Del-Zio trial, 155–56, 164, 165–66, 182–84, 189–90, 194; about fertility clinic in Virginia, 209; about first test tube babies, 141, 152, 156, 158–59, 170–73, 190, 212, 223; Shettles's, during Del-Zio trial, 180; Shettles's and Rorvik's, on sex selection, 53–58, 60, 67; and Shettles's IVF claims, 37–39, 181–82, 275n
Purdy, Jean, 172

Quadruplets, 236
Queens Federation of Churches (New York City), 206
Queen Victoria Medical Centre (Melbourne, Australia), 218
Quintuplets, 236

Rabbits, 2, 29–30, 33, 63, 130–31, 173–74, 187, 279n
Raëlians, 245–46
Ramsey, Paul, 71–74, 76, 77, 99, 104, 128
Randolph (Vermont), 180, 262, 289n
RBM Online, 264
Reagan, Ronald, 223
Recombinant DNA, 116–27. *See also* Genetic engineering
Reed, Candice Elizabeth, 212, 224, 238
Reed, Linda, 212
Regulation of medical research, 5; inadequate, 11–12, 46, 58, 76–77, 223, 253; involving human fetuses, 11–12, 80–82, 91, 111–12, 116, 128–32, 211, 231, 253–57; involving recombinant DNA, 123–24; and letters of assurance, 59, 95, 278n; mechanisms for, 46, 90–92, 124–26, 136, 193, 253–57; recommendations for, 90, 121, 135, 254; standard procedures of, 58–59; and Tuskegee Syphilis Study, 84, 283n. *See also* Bioethics; Consent forms; Government funding; Knowledge
Reproduction (assisted): fears about, 13–14, 27–28, 255, 257–59; IVF as most common technique of, 233; limiting of, 6–7, 10, 243; "rights" to, 243–44; and sexual intercourse, 6, 31, 52–53, 70, 135–36, 174, 243; social factors, 6–7, 10–12; statistics on, 236, 299n. *See also* Artificial insemination; Bioethics; Cloning; Fertilization (artificial); GIFT; In vitro fertilization; ZIFT
Reproduction (sexual), 6, 31, 52–53, 70, 135–36, 174, 243
Respirators (mechanical), 174
Riggs, Bobby, 105–6
"Right to life" movement. *See* Abortion: opponents of
Roberts, John D., 278n
Robinson, Angela, 289n
Rock, John, 30, 31, 51
Rocky Flats nuclear weapons plant, 157, 290n
Roe v. Wade decision, 6, 220
Rogers, Michael, 122
Rogers, Paul, 147–48
Rolling Stone, 122
Roosevelt Hospital (New York City), 168
Rorvik, David M.: background of, 147; on cloning, 144–47, 149, 246, 289–90nn; as Shettles's coauthor, 53–55, 60, 63, 67, 144, 147, 276n
Rose, Molly, 33, 39, 41, 43
Royal College of Obstetricians and Gynecologists (Great Britain), 114
Ruddle, Frank, 215–16, 257–58
Rugh, Roberts, 53
Rutherford, Ernest, 32
Ryan, Kenneth J., 128

Sadler, William, 90
Sagan, Carl, 201

St. Luke's Hospital (New York City), 167–68, 291n
Scandinavia, 87–88
Schweiker, Richard, 223
Science: concerns about dark side of, 7–8, 14–15, 65–77, 109–10, 116–32, 156–57, 257–60, 265–67; vs. dogma, 76; and fetal research moratorium, 18, 116, 128–32, 254–55; predictions about, 75, 257–59, 265–67; and recombinant DNA moratorium, 116–27; U.S.'s 1960s love affair with, 39–43; wonder of, 51, 79, 174. *See also* In vitro fertilization; Knowledge; Reproduction (assisted); Scientists; Slippery slope arguments; *and see specific biological sciences and discoveries*
Science, 30, 119, 126, 216
Scientific American, 71
Scientists: competency questions about, 58, 185, 187, 191–92, 213–14; IVF opposition from, 206; motivations of, 5, 12, 30, 32, 37–38, 182; and regulatory oversight, 12, 46, 76–77, 124, 136; as third party in IVF, 175. *See also* Regulation of medical research; *and see specific scientists*
Secondary infertility, 115
Second Sex, The (Beauvoir), 107
Secret, The (Hoffman), 251
Selzer, Richard, 256
Semen. *See* Sperm
Senate Health Subcommittee, 125–26
Sex selection, 53–54, 56, 60,144
Sex Surrogates, The (Rorvik), 147
Sexual reproduction. *See* Reproduction (sexual)
Shapiro, Harold, 255
Sheep (cloned), 248, 252, 255, 301n
Shelley, Mary Godwin, 1, 8–9, 100, 104–5, 284n
Shelley, Percy Bysshe, 8–9
Shettles, Landrum Brewer, 4, 5, 90, 157; on Australia's first test tube baby, 212; career of, 20–22, 35–39, 56–60, 63, 103, 134, 262–63; and cloning, 144–45, 262–63, 289n; and Del-Zios' IVF attempt, 3, 19, 23, 95–98, 101–2, 115, 160–61, 163–64, 169, 185, 187, 192; and Dennis, 116, 159, 180; and ethics, 22, 54, 63, 95, 97–100, 144–45; family of, 21, 57, 204, 263; IVF claims of, 35–37, 43, 181–82, 275nn; IVF research of, 19, 20–25, 35–39, 51–55, 181, 190–92; and lab vandalism, 102–3; personality and habits of, 20, 57, 68, 180–82, 204; photographs by, 20, 35, 51–53, 55, 60, 67, 178–79, 278n; and publicity, 37–39, 53–58, 60, 67, 180, 181–82, 275n; on sex selection, 53–54, 56, 60, 144; testimony regarding, in Del-Zio trial, 160–61, 163–64, 168–69, 177–78; as trial witness, 180–82; as urging Del-Zios' lawsuit, 115, 286n; Vande Wiele's relationship with, 20, 24–25, 54–60, 63, 144, 205, 230, 284n
Shriver, Eunice Kennedy, 45–47, 66, 88, 91
Shriver, Maria, 88
Shriver, Sargent, 45, 69
Singer, Daniel, 89
Sinsheimer, Robert, 119, 123
Situational ethics, 99–100
Six-Million Dollar Man, The (television show), 109–10
Skinner, B. F., 68
Sleeper (movie), 143
Slippery slope arguments, 10–11, 70, 75–76, 211–12, 215–16, 231–32, 237–38; and artificial insemination, 10, 28–29; and cloning, 10–11, 138, 249–50; Eunice Shriver's concerns about, 45–47; and IVF, 10–11, 72–74, 99–100, 135, 143–44, 265–66. *See also* Bioethics; Knowledge: fear of; Science: concerns about dark side of

Sloane Hospital for Women (New York City), 230
Society for Gynecological Investigation, 128
Society for Pediatric Research, 128
Sophocles, 71
Soupart, Pierre, 130–31, 136–37, 205, 223–24, 254, 297n
Soviet Union, 32, 273n
Space Age, 39, 157
Sperm: in artificial insemination, 27–29; capacitation of, 41–44, 279n; Carr's, 221; and cloning, 142; Del-Zio's, 19, 20; Edwards's, 78; and fertilization, 34, 273n; fertilization of egg by more than one, 34, 82, 202; giant, 262–63; gonorrhea's destruction of, 26–27; in IVF, 2, 35, 36, 71, 78, 113; and meiosis, 41; Shettles's, 35; tails of, 43; weak, 34, 240–41, 243, 300n. *See also* Eggs; Embryos; Fertilization; Gametes; Zygotes
Spina bifida, 238, 243
Steel, Samantha, 226
Stem cell research, 247–48, 255–56, 264, 300n
Steptoe, Patrick, 44–45, 228, 231, 242; death of, 263; Ethics Advisory Board's visit to, 188–89; family of, 44; and first test tube baby, 133–34, 140–41, 150–51, 170–72; government funding for, 61; IVF research of, 69, 71, 77, 113, 137, 181, 188–89, 202, 210, 229; as laparoscopy developer, 44–45, 71
Steptoe, Sheena, 44, 171, 172
Stern, Barbara Lang, 168
Stewart, Charles E., 158, 159, 165, 166, 168, 180, 182–83, 186; jury instructions of, 193–96
Stone Ridge Country Day School of the Sacred Heart (Maryland), 88
Styron, William, 68, 280n
Suicide (assisted), 11
Sullivan, Walter, 226
Surrogate motherhood, 50, 75, 135, 266
Sweeney, William: death of, 263; as Doris Del-Zio's infertility specialist, 1–5, 97, 101, 111, 115, 160–61, 169, 181; as trial witness, 167–69; as urging Del-Zio lawsuit, 115, 286n
Sydney (Australia), 217
Syphilis, 82–84, 87, 90

Tapley, Donald, 272n
Technology assessment, 70, 174–75, 292n, 293n
Teilhard de Chardin, Pierre, 47
Teresa, Mother, 68
Test tube babies: from artificial insemination, 26–29, 67; jokes about, 147, 189–90; number of, 226, 228. *See also* In vitro fertilization
"Test Tube Baby Is Coming, The" (Rorvik), 55
Three Mile Island nuclear power plant, 290n
Time, 146, 147
Time/CNN poll, 248
Times (London), 50
Today Show (television show), 264
Todd, Duane, 97
Tonight Show, The (television show), 147, 189–90
Toran-Allerand, Dominique, 23–25, 97, 102–3, 196
Total Woman, The (Morgan), 107–8
Toxemia, 150, 151
Transplantation: of fetal tissue, 254; of hearts, 10, 11, 37–38, 116, 227n; of other organs and bone marrow, 38, 47–48, 174. *See also* Implantation
Transposition heart defects, 238, 243
Triplets, 236, 237
Triploidy, 202, 294n, 295n
Tubal insufflation, 45
Tuomey, Theo, 88

Tuskegee Syphilis Study, 82–84, 87, 90
Twins: and "ensoulment," 81; first IVF, 218, 296n; number of, among IVF births, 237; risks associated with, 236. *See also* Multiple births

United States: first test tube baby in, 226; IVF multiple births in, 236–37; lack of federal IVF funding in, 11–12, 61–63, 79–80, 89–90, 130, 134–35, 205–6, 211, 223, 230; number of IVF procedures and clinics in, 230, 233; public opinion about reproductive technology in, 50. *See also* Congress; Department of Health, Education, and Welfare; Politicians; Regulation of medical research
University of Bologna (Italy), 31
University of Chicago, 137
University of Hull (England), 112
University of Leeds (England), 112, 114
University of Pennsylvania, 230
University of Sheffield (England), 114
University of Southern California, 230
University of Texas Medical Center, 230
University of Virginia (Charlottesville), 92

Valiant Ventures Ltd., 245
Vanderbilt University (Tennessee), 130, 136, 230
Vande Wiele, Betty, 24, 232
Vande Wiele, Jos, 25, 271n
Vande Wiele, Raymond, 157; attitudes toward IVF, 4, 58–59, 95–96, 120–21, 173, 185, 186, 214–15, 229–32; on bioethics panel, 89; career of, 24–25, 184–85, 204, 205, 297n; death of, 232, 263, 298n; and Del-Zios' IVF attempt, 24, 25, 95–98, 101–2, 115–16, 163, 166, 167, 177, 178, 185–88, 190, 194, 196, 197–98; Del-Zios' suit against, 115–16, 155–69, 177–96, 261; as director of fertility clinic, 230–31; and Shettles, 20, 24–25, 54–60, 63, 144, 205, 230, 284n; and Sweeney, 167–68, 291n; as trial witness, 184–88
Veeck, Lucinda, 227
Venereal diseases. *See* Gonorrhea; Syphilis
Vietnam War, 7, 119
Virginian-Pilot, 210, 213, 226–27
Virginia Society for Human Life, 206, 210

Wall Street Journal, 146
Walters, LeRoy, 89
"War on cancer," 61
Washington Post: on cloning, 70, 147; on "decapitated fetuses," 87–88; on infertility clinic in Virginia, 211–12; and IVF, 206; on Kennedy Center opening, 64; on test tube babies, 173, 228
Washington University (St. Louis), 240
Washkansky, Louis, 47
Watson, James, 64, 69; on cloning, 74–77, 170, 217; as DNA structure discoverer, 32, 74; on Edwards as "baby killer," 76, 77, 88; and genetic engineering, 119, 264–65, 287n, 302n
Weinberger, Caspar, 131–32
Wells, H. G., 258
Westinghouse Group, 261
"Whither Thou Goest" (Selzer), 256
Who Should Survive? (film), 65, 69
Wiesel, Elie, 68
Wiesenthal, Simon, 271n
Winston, Robert, 242, 299n, 300n
Woman's Doctor (Sweeney), 168, 291n
Woman's Medical Guide, The (Rorvik), 147
Wombs: artificial, 32, 211. *See also* Implantation
Women's Legal Defense Fund, 128
"Wonderful World of Chemistry, The" (1964 World's Fair), 39

Wood, Carl, 218, 224, 231, 297n
Wood, Judy, 218
World Medical Association, 58–59

X-ray, 45, 49

Yale University, 89, 215–16, 230, 257–58
Your Baby's Sex: Now You Can Choose (Shettles and Rorvik), 53–54, 144

Zero Population Growth (ZPG), 6–7, 108
Zervoudakis, Ionnis, 269n
ZIFT (zygote intrafallopian transfer), 234
Zona drilling, 234
Zona pellucida, 34, 41, 178
ZPG (Zero Population Growth), 6–7, 108
Zygote intrafallopian transfer (ZIFT), 234
Zygotes: in cloning, 13; definition of, 2, 250; in IVF, 2, 72, 78–79; and meiosis, 41. *See also* Blastocysts; Eggs (ova); Embryos; Fertilization; Morulas; Sperm